Mathematical Explorations for the Christian Thinker

Mathematical Explorations
for the
Christian Thinker

Jason VanBilliard

Copyright © 2014 Jason VanBilliard
All Rights Reserved.
ISBN-13: 978-1500795764
ISBN-10: 1500795763

Images in this text:
Uncited images used in this book are either in the public domain or created by the author and are released to the public domain. Other images with a URL are used under a Creative Commons ShareAlike License 2.0 and 3.0 and you are free: "to share (to copy, distribute and transmit the work)" and "to remix (to adapt the work)" under the following conditions: attribution -you must attribute the work in the manner specified in this work (but not in any way that suggests that they endorse you or your use of the work) and share alike - if you alter, transform, or build upon this work, you may distribute the resulting work only under the same or similar license to this one.

Cover art by
David Steininger

Table of Contents

Chapter 0: Biblically Informed Mathematics 1

Part 1: Shape

Chapter 1.1: Introduction to Sameness 11
Chapter 1.2: Topological Equivalence 23
Chapter 1.3: Dimensions 33
Chapter 1.4: Fractals 49
Chapter 1.5: Non-Euclidean Geometry 65
Chapter 1.6: The Golden Ratio 89
Chapter 1.7: Proof and Beauty 103

Part 2: Reason

Chapter 2.1: Logical Reasoning 121
Chapter 2.2: Inductive and Deductive Reasoning 137
Chapter 2.3: Cause and Correlation 147
Chapter 2.4: Deception in Statistics 159
Chapter 2.5: Voting 177
Chapter 2.6: Probability 185
Chapter 2.7: Complexity and Chaos 197

Part 3: Number

Chapter 3.1: Special Numbers 213
Chapter 3.2: Prime Numbers 223
Chapter 3.3: The Real Number Line, π, and e 237
Chapter 3.4: Counting Infinity 253
Chapter 3.5: Modular Arithmetic and Finite Groups 269
Chapter 3.6: Algebraic Reasoning 283
Chapter 3.7: Combinatorial Reasoning 295

Chapter ∞: The Beginning 317

Preface

The goals of this book are different than those of many other mathematics books in your experience. First, I intend to expand your understanding of the world of mathematics. The typical K-12 curriculum focuses on arithmetic, algebra, and Euclidean geometry. Some students do reach the Holy Grail of Calculus but that subject is still in the same vein as the algebra/geometry track. You may ask, "What else is out there?" This book will help answer that question.

Secondly, this book shows how mathematics connects to the rest of life. This statement may cause you to think that you will see a lot of mathematical applications. You will not. Instead, after we study each mathematical topic, we will consider how this topic helps inform our answers to the enduring questions: Who are we? What is the nature of reality? How do we know if something is true? What is good? What is beautiful? These questions and their related sub-questions have been part of the human experience from the dawn of human history.

Finally, you should know that this book is unapologetically biblical in its approach. A biblical view of mathematics requires one to love God and love their neighbors through the study of mathematics. The pragmatic, applied study of mathematics does help the Christian to love his neighbor; however, without examining the creative discovery involved in studying mathematics, we do not fully appreciate God and his creation. Hence, creativity, reflection, and biblical integration opportunities are abundant. You will have the opportunity to answer important worldview questions from a biblical perspective in every chapter.

Acknowledgments

I would like to thank Martha MacCullough, who introduced me to a model for meaningful integration across all content areas, Debbie MacCullough, who helped me to fully apply that model to the field of mathematics, Timothy Yoder, who compelled me to write this book, Brenda Mellon Ebersole, who provided a detailed critique of my writing, two decades of students, who refined my practice of simplifying complex ideas, and my wife and children, who sacrificed many family hours during the writing process. *Soli Deo Gloria.*

To the Student

Each chapter is divided into four major sections. First, it is important that you take time to think about and complete each **Warm-up Activity**. Each activity is designed to help to start you down the path of thinking about relevant information or skills. Skipping these activities will make it more difficult for you to understand the content.

In the second and more robust **Concept Development** section, you will experience a topic in-depth. This includes making connections within mathematics as well as links to other subjects. Read this section *slowly and carefully*. One challenge most people face when reading mathematics is that they expect to be able to read mathematics the way they read a novel. Mathematics is different. You need to *frequently pause* to reflect on what you do and do not understand. You may ask, "Can I give an example and non-example of that idea?", "Do I need to reread that section?", or "Could I explain that to someone else?" It is important to reflect as you read.

The third section, **Something to Think About,** provides you with compelling questions that relate the mathematics you just learned to enduring questions. These questions may not be easy to answer; they are worthy of reflection and dialogue. You will continue to think about these types of questions long after finishing this book.

Each chapter ends with **Covering the Reading** questions and **Problems.** The Covering the Reading questions are designed to help you process what you have read. If you are unable to answer these questions, you should probably do some rereading. The expectations on the Problems are completely different. In mathematics, the difference between an exercise and a problem is that an exercise is something you already know how to do but should practice to gain mastery. A problem is something that you may not know how to do. In this section you will need to problem-solve. You may need to research. You may need to spend extra time reflecting. You may have to leave the problems and come back to the problems when your mind is refreshed. Taking time to reflect and do problems is usually necessary to fully understand what you have read.

To the Instructor

Unlike many mathematics books that serve as a reference, this book is meant to be read by the student. However, the methods for using this book heavily depend on your students reading comprehension skills, independent thinking skills, and study habits. I recommend three potential approaches.

Method 1: High-level student responsibility. The students read the text before class and complete the Covering the Reading questions. Class time is used to work with the teacher and small groups on questions about the subject and a review of the Covering the Reading questions. This may require a short lecture by the teacher related to a certain portion of the material. In addition, the teacher can reflect on her experiences with the topic. Students then do the Problems for homework and bring their work to the following class. Time is then spent discussing the problems as a class. This approach requires that the students be conscientious and that the teacher be flexible with the topics at hand.

Method 2: Mixed responsibility. The students read the text before class and complete the Covering the Reading questions. At the start of class, the teacher re-presents the material in his own way and incorporates the Covering the Reading into the interactive lesson. The teacher can then either assign the Problems for homework or pursue them as a class at the next meeting. This approach allows the teacher to present material in a different way than the book, one that may be more appropriate for a specific group of students. Essentially, the students would get the information in two different ways from two different people.

Method 3: High-level teacher responsibility. The teacher thoroughly presents the material in class. Students use the book as a resource while completing the Covering the Reading for homework or during the next class. Problems are handled the same way. This approach gives the teacher more control of the direction and pace of the course.

Each of these methods has its strengths and weaknesses. The way you use this book should be a function of who you and your students are. You may envision another way to use this book that is better than these three methods. Have fun teaching. I also hope that I have raised some questions you have never considered.

Chapter 0:
Biblically Informed Mathematics

Humanity has forever pursued the "enduring questions." They have also been called "worldview issues" and "essential questions." The heart of this preoccupation is that humanity has been wrestling with the questions of existence, truth, purpose, beauty, and ethics from the dawn of time. Who am I? How did we get here? Is there a God? What is really real? How do we know if something is true? What is truth? Is there a basis for good and evil? What is beauty? Who determines what is true, good, and beautiful?

The School of Athens- Raphael

Among other tools, theology, reason, and the study of the natural world have helped us to begin to answer these enduring questions. These various sources have led us to myriad competing views on answers to these questions. Thinkers have found that different academic subject help us tease out answers to many of the enduring questions.

Although each discipline aids in addressing these big questions, each one also grapples with its own particular sub-questions. For example, history and the social sciences confront the questions, "Who is the human?" and "Who controls the historical narrative?" Science asks, "What questions can science address and what questions are not accessible through science?" and "How do we determine what is true?" Music forces us to contemplate, "What constitutes beautiful music?" and "Are there eternal rules of order or is improvisation at the heart of music?" In the field of art we ask, "Who decides what is and is not art?" and "Must beauty and art overlap or can they be independent of each other?" Literature asks if meaning comes from the author or from what the reader perceives in the text. The world of business asks how you can place value on people, goods and services. Every subject deals with its own questions that help us answer the enduring questions.

Warm-up 1

What kind of enduring questions do you anticipate thinking about in this book? How does mathematics help to answer the enduring questions? Try to list 3 questions that you believe relate to mathematics.

1. _____
2. _____
3. _____

Concept Development 1

The field of mathematics tends to wrestle with questions about the nature of reality and evaluating how we determine if something is true. However, it also branches into the nature and purpose of the human being, discussions of beauty, and even ethical considerations. In addition to many others, mathematical philosophers have asked these questions: What is mathematics? What is number? Is it a noun or an adjective? What is infinity? Does it exist? What does it mean for something to *exist* mathematically? What are the limits of mathematics? How much math is "out there"? Is mathematics discovered or created? Can anyone do math? How does math relate to science? Is math part of nature? Is there randomness in the universe? Can everything be described mathematically?

Warm-up 2

Regardless of people's religion or creed, everyone must confront these types of enduring questions if they are serious about the foundations of mathematics. However, in this book we seek to pose and address the questions from a Christian perspective. We submit the questions to the scrutiny of the Bible and seek to understand these issues under the lordship of Christ.

Oftentimes, the way we ask a question changes how we think about its answer, so we need to carefully examine the questions themselves. How could we contextualize the following questions to be answered from a more specifically biblical perspective. Rewrite these questions to explicitly include a theistic twist. However, do not worry about answering the questions now.

1. What is mathematics?
2. Is mathematics discovered or created?
3. Can anyone do math?
4. Is math part of nature?
5. Is there randomness in the universe?

Concept Development 2

Let's begin the journey toward biblical perspectives on mathematics with an example. As we will do throughout this book, we will now examine a mathematical topic that expands our understanding of the world of mathematics. The question we will consider here is the relationship between God, mathematics, and nature: within this context we will address related enduring questions. This section will introduce the Fibonacci sequence, which this book will later present more exhaustively. The purpose for the introduction here is to illustrate how we can pursue biblical answers to the enduring questions related through mathematics.

The *Fibonacci sequence*, named after the Italian mathematician Fibonacci of Pisa, is historically and typically presented as a recursively defined formula. A *recursive* formula is one defined by previous values while an *explicit* formula produces the value simply by calculating the desired term. For example, the sequence $30, 40, 50, 60, 70, 80$ ... can be expressed explicitly as $s_n = 10n + 20$. That is, the 5th element of the sequence, s_5, is the same as ten times the index, n, plus 20; $10 \times 5 + 20 = 70$. This sequence can also be expressed recursively. Notice that each new term can be found by adding 10 to the previous term. In addition, the first term is 30. Recursively, the sequence is

$$\begin{cases} s_n = s_{n-1} + 10 \\ s_1 = 30 \end{cases}$$

The Fibonacci sequence is similarly defined recursively. The sequence is

$$1, 1, 2, 3, 5, 8, 13, 21, 34, \ldots$$

Notice that each term is the sum of the two previous terms.

$$5 = 3 + 2$$
$$34 = 21 + 13$$

Recursively, the Fibonacci sequence is the sum of the two previous terms such that the first two terms are 1 and 1.

$$\begin{cases} F_n = F_{n-1} + F_{n-2} \\ F_2 = 1 \\ F_1 = 1 \end{cases}$$

The first curiosity we come across in this book is that the Fibonacci numbers, $1, 1, 2, 3, 5, 8, 13, 21, 34, \ldots$ are abundant in flowers and plants: Pine Cone (8), Black Eyed Susan (5), Trillium (3), Shasta Daisy (21):

How do we explain this curious frequency?

Something to Think About

We need to step back and think about the relationship among God, mathematics, and nature, but first let's take a look at how people have explained the math-nature relationship without God. One view as to why the Fibonacci numbers appear frequently in nature contends that the Fibonacci numbers and nature are *not* explicitly linked. Fibonacci in nature is not a law but a prevalent tendency; other sequences are also found in nature. Fibonacci just happens to be frequent.

Another prevalent view is that things naturally grow in Fibonacci numbers because Fibonacci is most efficient. From an evolutionary standpoint, efficient and effective growth in a plan enhances its chances of survival. The optimal pattern for growth must be a compact and efficient layout in which existing, complete structures can all reproduce. If we continue to study the Fibonacci numbers and the related rectangles and spirals, we can see that these conditions are met in Fibonacci. This is an evolutionary explanation for Fibonacci in nature.

However, we seek to more completely understand the universe from a perspective informed by both the general revelation of nature and the special revelation of the Bible. What is a biblical, theistic perspective on the Fibonacci numbers and nature? We can pursue a related question even if the Fibonacci numbers are not truly in nature, as the first argument proposes. We can also re-pose the question as a more general, enduring question regarding the relationship among God, mathematics, and nature. Furthermore, since we can only study God, mathematics, and nature through our own human lens, we can ask what the relationship is among God, humanity, mathematics, and nature.

Covering the Reading

1. What do you believe are the three most important enduring questions? Why?

2. Complete "Warm-up 2": Rewrite these questions to *more specifically* include a theistic twist.
   A. What is mathematics?
   B. Is mathematics discovered or created?

C. Can anyone do math?
D. Is math part of nature?
E. Is there randomness in the universe?

3. Write the next 5 terms of the Fibonacci sequence: 1, 1, 2, 3, 5, 8, 13, 21, 34, ...

4. Can you think of something in nature that you have seen recently that has 3, 5, or 8 of something?

Problems

5 – 9. (a) Restate each of the following quotes in your own words, (b) explain how each quote does or does not relate God, nature, humanity, and mathematics, and (c) indicate your level of agreement with the quote.

5. Galileo (astronomer, scientist, mathematician): "Mathematics is the language with which God has written the universe."

6. Gauss (one of the three greatest mathematicians of all time): "Number is merely the product of our mind."

7. Kline (math historian): "Mathematics is a human activity and is subject to all the foibles and frailties of humans."

8. Einstein: "How is it possible that mathematics, a product of human thought that is independent of experience, fits so excellently the objects of physical reality?"

9. Davis & Hersh (they have written about various philosophies of mathematics): "...the whole of mathematics exists externally,

independently of man, and the job of the mathematician is to discover these truths."

10. Describe your view of the relationship among God, nature, humanity, and mathematics. What does the Bible have to say about these questions?

Part 1
Shape

Chapter 1.1
Introduction to Sameness

Mathematics could be considered a study in "sameness." The tools developed in mathematics are frequently used to solve real-world problems by *modeling* a real-world situation. In problem solving, we often look at a situation and see a simplified mathematical version of it; the two are "the same" even though the situations are not identical.

From our very first introduction to mathematics, we are asked to move from concrete to abstract thinking; we are asked, "What is 2 apples and 2 apples?" Our classrooms are usually void of apples so we draw pictures of two "apples" and another two "apples." Are we actually adding apples? We begin by adding pictures of apples which are *the same enough* to do the job. From here we progress in abstraction so that any time we are asked to add 2 of any object to 2 of the same object we are comfortable saying there are 4 of that object without ever touching or using those objects. We proceed to make the concept fully abstract by generalizing $2 + 2 = 4$ and use this property to solve a variety of problems. Why can we do this? Because "2" is used to represent anything that is the same as this many " * * " of anything.

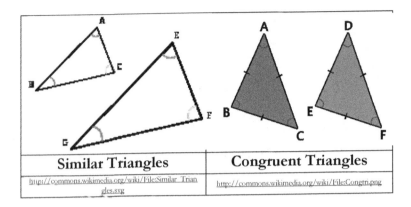

Similar Triangles	Congruent Triangles
http://commons.wikimedia.org/wiki/File:Similar_Triangles.svg	http://commons.wikimedia.org/wiki/File:Congtri.png

This sameness extends past arithmetic to algebra, geometry, and beyond. In geometry we use sameness in a variety of ways; every triangle is "the same" in certain ways. Similar triangles are "the same" in certain ways. Congruent triangles are "the same" in more ways, but not all ways (two triangles can be congruent even if they are in different orientations; one can be upside-down from the other yet they are congruent). Sameness can come in different degrees.

Warm-up Activity

Think of 10 instances (5 in mathematics, 5 beyond mathematics) when you use the idea of 2 things being "the same" without them being identical. In each case, how are they the same?

Things that are "the same"	Sameness characteristics
1.	
2.	
3.	
4.	
5.	
6.	
7.	

8.	
9.	
10.	

Practicing Sameness: Investigating Platonic and Archimedean Solids

A *regular polygon* is a polygon in which all of the sides are congruent and all of the angles are congruent. A STOP sign is in the shape of a regular octagon. The headquarters for the US armed forces is in the shape of a regular pentagon.

Similarly, a *regular polyhedron* is a 3-dimensional object that is analogous to a regular polygon. It is made up of identical polygons which are related to each other in equal angle measures. There is a very technical definition for regular polyhedron, but for now we only need to look at them. There are exactly five regular polyhedrons. They are known as the 5 Platonic Solids.

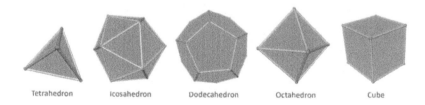

Tetrahedron   Icosahedron   Dodecahedron   Octahedron   Cube

Plato

From Raphael's *School of Athens*

Each Platonic Solid is made up of vertices, edges, and faces. The cube (or hexahedron) has 8 vertices (corners), 12 edges, and 6 faces. We get the name "hexahedron" from the cube's 6 faces. Try to complete the chart for the other polyhedrons. You may need to do some research to complete the chart.

Name	Vertices	Edges	Faces
Tetrahedron	4	6	4
Cube (Hexahedron)	8	12	6
Octahedron	6	12	8
Dodecahedron	20	30	12
Icosahedron	12	30	20

In addition to the 5 regular Platonic Solids, there are 13 - and only 13 - semi-regular Archimedean Solids. Archimedean Solids consist of more than one regular polygon but have the same pattern of polygons around each vertex. The solids have what seem to be funny names at first, but each name reveals some of its characteristics. It is worth trying to pronounce each one.

Archimedean Solids

truncated tetrahedron		snub cube	
cuboctahedron		icosidodecahedron	
truncated cube or truncated hexahedron		truncated dodecahedron	
truncated octahedron		truncated icosahedron	
small rhombicuboctahedron		small rhombicosidodecahedron	
great rhombicuboctahedron		great rhombicosidodecahedron	

Robert Webb's <u>Stella software</u> created these images: http://www.software3d.com/Stella.php		snub dodecahedron	

Practicing Sameness

Take some time to examine the 5 Platonic Solids and the 13 Archimedean Solids. In a certain way, the 5 Platonic Solids are "the same" and the 13 Archimedean Solids are "the same." They fall into two different categories. Practice trying to sort the 18 solids into two different categories. Can you classify them into three different categories? Four different categories? How would you describe your categories? Are the categories overlapping or mutually exclusive (a solid cannot be in more than one category?

More on Platonic and Archimedean Solids

There are relationships between the Platonic Solids and Archimedean Solids as well as among the Archimedean Solids themselves. For example, if we were to shave the right amount off of each of the corners of a cube, we arrive at the truncated cube. If we continue to truncate, we arrive at the cuboctahedron.

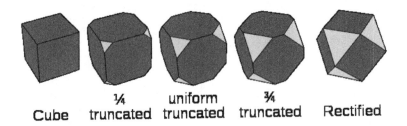

Cube — ¼ truncated — uniform truncated — ¾ truncated — Rectified

If we continued to truncate, we would arrive at the truncated octahedron. Finally, completing the morph we would arrive at the cube's *dual*: the octahedron. Cubes and octahedrons are duals:

the midpoints of the faces of one serve as the vertices of the other. Thus, the cube has 6 faces and 8 vertices while the octahedron has 8 faces and 6 vertices.

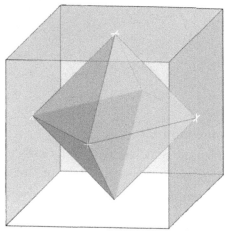

http://en.wikipedia.org/wiki/File:Dual_Cube-Octahedron.svg

In the progression between the two solids of the dual there are 3 Archimedean Solids.

Cube → Truncated Cube → Cubeoctahedron → Truncated Octahedron → Octahedron

A similar relationship exists between the duals dodecahedron and icosahedron: Dodecahedron → Truncated Dodecahedron → Icosadodecahedron → Truncated Icosahedron → Icosahedron.

The other Archimedean Solids are derived from various shavings or truncations of certain Archimedean Solids. For example, the great and small rhombicosadodecahedrons can be found by truncating the icosadodecahedron.

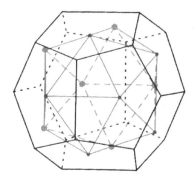

Historical Connection

The 5 Platonic Solids were used by early philosophers and later by scientists to explain the structure or content of the universe. For example, Plato saw each of the 5 as constituent parts of the universe:

Tetrahedron	Fire
Cube	Earth
Octahedron	Air
Dodecahedron	The Aether (universe)
Icosahedron	Water

Later, Johannes Kepler used the 5 Platonic Solids as an analogy of the Five Worlds and the harmony of the universe.

Fig. 37 —Kepler's Analogy of the Five Solids and the Five Worlds.

Something to Think About

Do you think some of the Platonic and Archimedean solids are nicer than the others? Do you like some more than the others? Would other people agree with you? Is there such a thing as a "nice shape"? An "ugly shape"?

Covering the Reading

1. How has your concept of "sameness" been expanded as a result of the reading?

2. Complete the table:

Name	Vertices	Edges	Faces
Tetrahedron			4
Cube (Hexahedron)	8	12	6
Octahedron			8
Dodecahedron			12
Icosahedron			20

3. What is the major difference between the Platonic and Archimedean Solids?

4. Categorize the 18 solids into two or three different categories. Describe the categories by identifying what is the same about all of the solids in each category.

5. What is a truncated dodecahedron? How is it different than a regular dodecadedron?

6. What is the relationship between a dodecahedron and an icosahedron?

Problems

7. Consider the table in #2 above. Can you find the mathematical relationship between V (the number of vertices), E (the number of edges), and F (the number of faces)? Does this relationship hold for the Archimedean Solids?

8. Explain why the dual of a tetrahedron is a tetrahedron.

9. The *net* of a 3-dimensional solid is the unfolded two-dimensional version of that solid. Alternatively, a net can be folded into a solid: for example, a cube has a net made of six squares. Match each solid to its net.

	Solid		Net
1.		A.	
2.		B.	
3.		C.	
4.		D.	

10. In this chapter we saw how things can be categorized as "the same" even if they are not identical. Do you believe that humans innately categorize? Do we create the categories or are they already in the fabric of the universe and we discover them? On what basis do you come to your conclusions? Does your thinking include biblical perspective on how God made humans to interact with His world?

11. Look at the 18 solids. Do you have a favorite solid? Are there solids that you like more than others? Can we objectively say one is nicer or better than another? Is there any basis for this type of judgment? Can we EVER say ANY shape is nicer than another?

Chapter 1.2
Topological Equivalence

Topological equivalence is a type of sameness. In this chapter topological equivalence will *not* be defined for you. Instead, you will grow in an understanding of this type of sameness and gradually develop your own definition.

Historical Context

One of the most famous topology problems in math history is the Konigsberg Bridge Problem posed by Leonhard Euler in 1735. The Prussian city of Königsberg is separated by a river; there are also two islands in that river. Seven bridges connect the four land masses as pictured below.

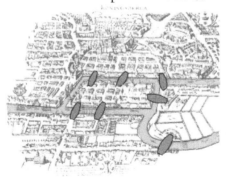

http://en.wikipedia.org/wiki/File:Konigsberg_bridges.png

Euler's famous problem asks us to walk over all seven bridges without repeating a bridge. You may start anywhere and end anywhere, but you must cross every bridge exactly once. Try the puzzle a few times. See if you can do it. (For you lateral thinkers out there- no, you may not swim, fly, teleport, etc… just take the puzzle for what it is!).

Problems like this can be simplified into a *graph*. This is not a graph like you may encounter in Algebra 1. Instead, this graph is made up of *edges* and *vertices* (also known as points or nodes). If you let a vertex represent a landmass and an edge represent a bridge, then the Königsberg problem simplifies to *traversing* this graph:

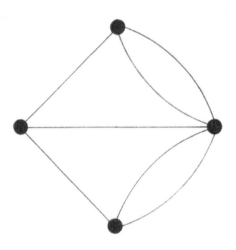

An extensive analysis of this type of graph was first attributed to Leonhard Euler, who was born in 1707. Euler will be frequently mentioned in this book as he contributed to nearly every field of mathematics. Euler (pronounced "oiler") is easily the most notable significant contributor to the field of mathematics who was also a professing follower of Christ.

While considering the Königsburg Bridge Problem, Euler pursued the question, "Under what conditions is a graph traversable?" What makes this question significant is that many situations can be modeled by graph theory. Water pipe flow, mail routes, and traffic patterns can all be simplified to a graph problem. What makes graphs so powerful is that, although they are not identical to the problem they are *the same enough* that they reveal the essential relationships between places.

In geometric topology, we try to identify if two objects are *the same enough* to be considered topologically equivalent. The Königsberg Bridge Problem is topologically equivalent to its graph. What makes two objects topologically equivalent?

Warm-Up Activity

Mathematicians joke that a topologist cannot tell the difference between a donut and a coffee mug.

1. Besides the perfect way to start a day, how is a donut *exactly* like a coffee mug?

2. List 10 other "donuts."

Developing the Concept

The donut/coffee mug example is a 3-dimensional topology example. In this section we are focusing on 2-dimensional objects that are topologically equivalent. In each case below you will see four objects in a set. Three of the objects in the set are topologically equivalent, one of them is not. Try to determine which one is not equivalent, and then describe what makes the other three equivalent.

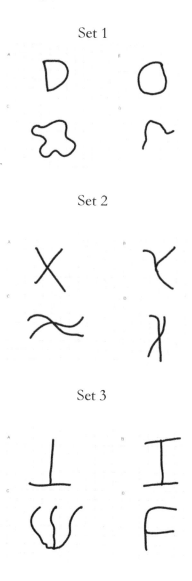

Set 4

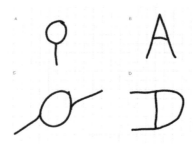

Which did you choose? Why? In set 1, D is topologically different. Notice that A, B, and C are all "loops" while D is just a string. What does this mean for your definition of topological equivalence? Things can be straight, curved, or wiggly and topological equivalence is not impacted. In set 2, A, C, and D are topologically equivalent. What is different about B? In A, C, and D, four segments meet at the intersection while in B only three segments meet at the intersection. Set 3 and set 4 are a bit more complex. In set 3, B, C, and D are topologically equivalent. Can you explain why? In set 4, A, C, and D are topologically equivalent. Why?

> **"What is topological equivalence?"**
> Try to write a definition based on your experiences so far in this chapter.

Let's increase the complexity of the activity:

Set 5

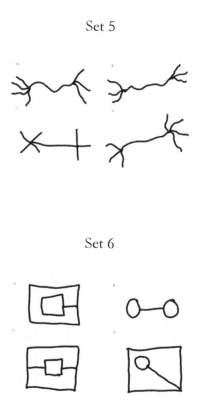

Set 6

Can you provide an argument for set 5 and set 6? Another way to think about topological equivalence is to pretend that you are an ant walking on these objects where the segments and curves represent pathways. What kind of options you have the intersection points? When would you loop back to the place where you started? When you hit a dead end? Considering these sets "as an ant" helps us to figure out how objects are connected.

In the chart below, draw an object that is topologically equivalent to each pair of letters:

Letters:	Topologically equivalent drawing:
W I	
A R	
T E	
e P	
p a	

Return to your definition of topological equivalence. How would you add to or change your definition? Hopefully you are developing an intuitive sense of what it means for things to be topologically equivalent. Try to take some time to modify your definition.

Something to Think About

Over the course of this chapter you built a new understanding of what was probably a totally new construct for you. You were "doing math." What does it mean to do math? To what extent were you creating? To what extent were you discovering? You were slowly *developing* an understanding of the mathematics. It is likely that your understanding was not simply an on/off switch. Are there levels of understanding in mathematics? Are there depths of understanding?

Covering the Reading

1. Can you solve the Königsburg Bridge Problem?

2. List 10 more "donuts" (topologically speaking).

3. List your answers for Set 5 and Set 6. Provide a rationale.

4. Sketch your topologically equivalent drawings to the letters in the chart above.

5. How do you define "topologically equivalence"? Would your definition apply to 3-dimensional objects as well?

Problems

6. In the chapter you thought about a "donut" with one hole. In mathematics it is called a torus. Consider a donut with two holes: a double torus. List 5 items topologically equivalent to a double torus. List 5 items topologically equivalent to a triple torus.

7. Group the following Greek letters into topologically equivalent sets:

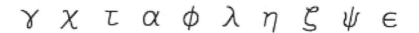

8. Group the following objects into topologically equivalent sets:

9. The following pair of graphs are not topologically equivalent. Explain *why* they are not topologically equivalent and then describe a change to the left image which will make the two figures topologically equivalent.

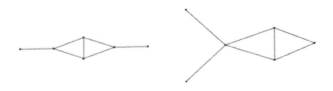

10. Describe a situation that can be modeled by a topologically equivalent graph (like the Königsburg Bridge Problem). Provide an example and explain how it could be useful.

11. Write the capital letters of the alphabet *in your handwriting* and group them topologically. Compare your grouping to someone else and identify differences.

12. In this chapter you were "doing math" by categorizing and sorting. When we "do math" do you believe we are creating or discovering? What is the basis for your response? Include the concept of topological equivalence in your answer. Did God make the categories or did we?

Chapter 1.3
Dimensions

What is a "dimension"? You can look the word up in a dictionary, but that definition will probably not provide much mathematical insight. In this chapter, we will experience and play with *dimension* in an effort to develop a deeper, intuitive understanding of dimension that goes beyond the dictionary definition. You will accomplish this by considering dimension from a variety of perspectives and then working to extend your understanding of dimension to four dimensions, five dimensions, and beyond. As we think through and pass three dimensions, we will rely on analogy and pattern to make inferences.

Warm-up Activity

Group each word from the "Word Bank," into categories with "Point," "Line," "Plane," and "Space."

Word Bank: square, 3-dimensional, location, area, volume, cm, pyramid, cm^2, cube, length, 2-dimensional, 0-dimensional, cm^3, 1-dimensional, line segment, triangle.

Point	Line	Plane	Space

Concept Development:

You have begun to think about what dimensions are by considering related geometric objects. Now we will consider what it takes to expand from one to two dimensions and from two to three dimensions. Try this mental exercise.

Begin with a point. A point is said to be zero-dimensional. It has only location.

Drag the point down one inch.

We have traced a one-dimensional line segment. It has length and can be measured linearly, say in inches or centimeters. If we continue dragging in the same direction, we exist in that one-dimension; we will stay only on that line. What would it take to move to a second dimension?

Now drag the segment to the right one inch.

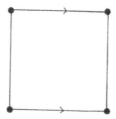

We have now traced a two-dimensional square. This allows us to move up and down as well as left and right; we now exist in a two-dimensional plane. We can consider location, length, and now *area*. The area might be measured in square inches (in^2) or square centimeters (cm^2). What did it take to move from one dimension to two dimensions? We had to move in a completely new direction. We chose for that direction to be perpendicular to the original segment. However, it could have been any new direction *except* parallel. Moving parallel to the original segment does not move us into a new dimension.

We can continue to exist in two dimensions and move up/down or left/right or up to the left or down to the right or even 10 degrees to the right of up. All of those directions keep us in the same plane; we stay in the same two dimensional surface.

If that is the case, then how can we move to three dimensions? Up/down? Left/right? Where is the third dimension? We would have to move in a completely different direction – a direction that is *independent* of the other two directions. But where is that dimension? Up to the right is *dependent* on the up/down, left/right construct. 10 degrees to the right of up is also *dependent* on the up/down, left/right construct. Yet even though we know these facts, we can still move to a third dimension. We just have to move differently.

Try this: slide the entire two-dimensional square frontwards or backwards by an inch. We have now created a three dimensional cube (illustrated on this two dimensional paper).

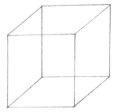

At this point we want to stop and consider our original question, "What is a dimension?" Try to write a working definition of a dimension.

Drawing Objects

Now, let's draw some objects using this dimensional concept. Using these drawings, we will then try to extend our understanding to four dimensions. Our first goal is to draw a four-dimensional "cube" which is called a hypercube or a tesseract. In order to make your sketches clear, use a different color for each dimension. Use one color for the up/down dimension and a different color for the left/right dimension, and so on.

This time, let's begin by connecting two 0-dimensional points to make a segment.

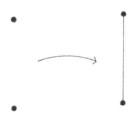

Now, connect two 1-dimensional segments to make a square.

Next, connect two 2-dimensional squares to make a cube.

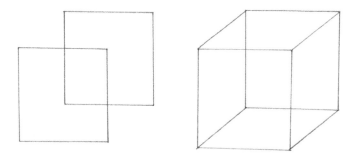

What is the next dimensional step? Connect two 3-dimensional cubes to make a hypercube.

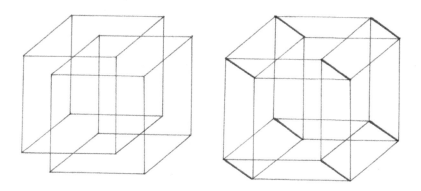

This object is hard for us to imagine because we cannot experience it. However, mathematical properties can be extended into more than three dimensions.

Recall, each object is made up of lower dimensional objects. For example, a cube contains vertices (the points on the corners), edges (the segments connecting vertices), and faces (the segments connecting edges). Similarly, a hypercube will contain all of the lower dimensional objects including cubes, also referred to as cells. Consider the chart below. First, verify the number of vertices, edges, and faces in each object. Using this information, can you predict the properties of the hypercube? Use mathematical reasoning and your drawings to help you count.

Object	Dimensions	Vertices	Edges	Faces	Cubes (Cells)
Point	0	1	0	0	0
Segment	1	2	1	0	0
Square	2	4	4	1	0
Cube	3	8	12	6	1
Hypercube (tesseract)	4	?	?	?	?

Probably the most difficult thing to see is the number of cubes (or cells) in the hypercube. To understand this, let's use an analogy. The cube has 6 faces. These 6 faces can be seen by removing 1 dimension at a time:

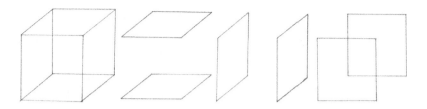

Using this method, try to determine the number of cubes (cells) in a hypercube. For instance, below are two cubes (cells) resulting from removing one of the dimensions. Using this method, you can probably be certain of the number of edges and faces as well

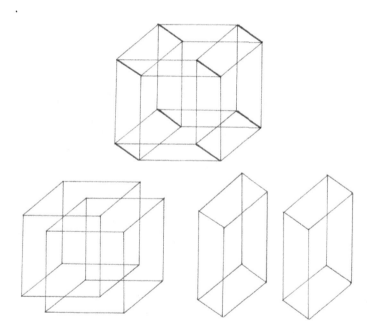

Through careful and creative analysis, we are able to make statements about objects that only exist in mathematical reality. Yet, the mathematical properties of objects beyond our three-dimensional experience find application in our world.

Subject-to-Subject Connection: Literature and Art

Higher dimensional concepts can be found in both literature and art. Salvador Dali's 1954 oil-on-canvas *Crucifixion (Corpus Hypercubus)* shows Christ crucified on an *"unfolded"* hypercube. To understand the painting, consider that a a 3-dimensional cube unfolds into a 2-dimensional cross made of 6 squares.

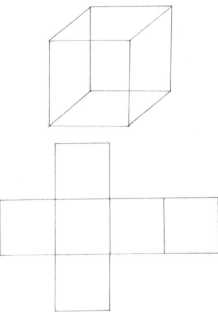

Similarly, a hypercube unfolds into 8-cubes.

In the 1962 adolescent literature sci-fi/fantasy classic *A Wrinkle in Time*, author Madeleine L'Engle has her character travel from one 3-dimensional location in space to another 3-dimensional location by "tessering" out into the 4th dimension and returning into regular space. If you find this to be a difficult concept, L'Engle anticipated your objection: she explains that it is much easier for young children to understand this concept because they are not stifled by adults' long-term experiences with 3-dimensions.

Mathematical Explorations | 43

Edwin Abbott's 1884 satire entitled *Flatland: A Romance of Many Dimensions* depicts the 2-dimensional world of Flatland in which people are 2-dimensional shapes moving in a 2-dimensional plane. The book becomes mathematically interesting when a 3-dimensional sphere visits Flatland and tries to explain the third dimension to 2-dimensional people. This struggle is well illustrated in the 2007 adaptation *Flatland: The Movie* voiced by Martin Sheen, Kristen Bell, and Tony Hale. As you can see from these examples, authors and artist have long been considering the implications of mathematical concepts in other areas of thought: literature, painting, and film.

Historical Connection

The mathematics of the ancient Greek mathematicians was geometry. Late Greek mathematicians began to develop algebra, but that discipline matured in the Middle East during the West's "Dark Ages." Greek geometry was rooted in real objects. A number represented either a discrete number of things or a length. Adding two numbers would mean adding two lengths. You may think of it as $3 + 5 = 8$ was understood as 3 inches plus 5 inches equals 8 inches. Multiplying two numbers meant area. $3 \times 5 = 15$ is the 2-dimensional area of a 3 x 5 rectangle. A resulting formula might be $A = l \times w$. This formula has meaning. Multiplying three variable values together would be a 3-dimensional volume: $V = l \times w \times h$. Multiplying four variable values together would have no physical meaning. Hence, a formula, or way of solving a geometric problem, would have no meaning if the formula involved multiplying four variable values together. It would be out of the realm of "mathematics" as it was understood at the time.

In the late Greek period, a disturbing thing happened. It would not be disturbing to us today as we algebraically work with equations involving "x^4." However, to the Greek mathematician who was experiencing the early stages of mathematical development, Heron's Formula was upsetting. Heron's Formula is used to find the area of any triangle. Given a triangle with sides length a, b, and c, you first calculate the semi-perimeter: half of the perimeter or $s = \frac{a+b+c}{2}$. Then Heron's Formula finds the area by calculating: $T = \sqrt{s(s-a)(s-b)(s-c)}$.

This formula may not appear to be troubling to us, but let us consider what the formula does. Under the radical, the "square root" sign, we multiply four values together. To the Greek, we are going out to the *unreal* fourth dimension. Then we take the square root of the 4-dimensional object to find a 2-dimensional area. Algebraically, this is not hard for us. However, to the Greek, we are basically "tessering"! We are going out into the 4th dimension to solve a 2-dimensional problem!

Something to Think About

Edwin Abbot, author of *Flatland: A Romance of Many Dimensions*, was a 19th century theologian. Many argue that the 3-dimensional sphere visiting 2-dimensional Flatland is a picture of a higher dimensional Jesus visiting and appearing as a lower-dimensional being. Whether or not that was the author's intent, the analogy is worth considering. When Sphere visits Flatland, he appears as a circle to the Flatlanders since a 2-dimensional slice of a sphere is a circle. He needs to explain the 3-dimensional world to those with no experiences of it. Similarly, Christ uses parables and analogies to communicate the nature of the Kingdom of God to us humans: "The kingdom of heaven is like…" Certainly, we don't want to limit God to a 4-dimensional being, but consider the challenges in trying to explain a spiritual "dimension" to people who believe only their 3-dimensional observations: *empiricists*. How can we explain the spiritual world to an empirically dominated society? Also, is it accurate/biblical/theologically responsible to consider God as a higher dimensional being as developed in this chapter?

Covering the Reading

1. In the warm up activity you interacted with objects of differing dimensions. List a 1-dimensional, 2-dimensional, and 3-dimensional object *not* identified in the reading.

2. a) Complete the chart:

Object	Dimensions	Vertices	Edges	Faces	Cubes (Cells)
Point	0	1	0	0	0
Segment	1	2	1	0	0
Square	2	4	4	1	0
Cube	3	8	12	6	1
Hypercube (tesseract)	4	?	?	?	?

b) *Explain* how you arrived at each number.

3. What is your working definition of "dimension"?

4. Pretend you are Sphere in Flatland. How would you explain a 3-dimensional world to someone from a 2-dimensional world?

5. Use Heron's Formula to find the area of a triangle with side lengths 4, 6, and 8.

Problems

6. Extend the chart to the 5-dimensional, square-based "penteract."

Object	Dimensions	Vertices	Edges	Faces	Cubes (Cells)	Hypercubes
Point	0	1	0	0	0	0
Segment	1	2	1	0	0	0
Square	2	4	4	1	0	0
Cube	3	8	12	6	1	0
Hypercube (tesseract)	4	?	?	?	?	1
Super-hypercube (penteract)						

7. The square-cube family studied in this lesson was based on constructing a parallel object and connecting both objects through an additional dimension. However, you could also build a triangle-tetrahedron family of objects. This is accomplished by connecting the previous object to a *point* out in the new dimension:

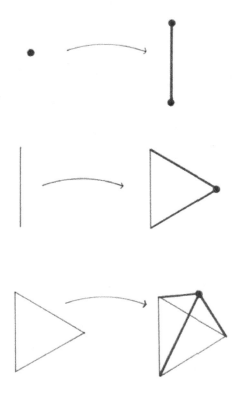

Now use this knowledge to complete the following chart:

Object	Dimensions	Vertices	Edges	Faces	Tetrahedrons (Cells)
Point	0				
Segment	1				
Triangle	2				
Tetrahedron	3				
Hypertetrahedron (pentatope)	4				

8. *Flatland: A Romance of Many Dimensions* - When Sphere visits Flatland, he appears as a circle to the Flatlanders (a 2-dimensional slice of a sphere is a circle). He needs to explain the 3-dimensional world to those with no experiences related to it. How can we explain the spiritual world to an empirically dominated society?

9. Is it accurate/biblical/theologically responsible to consider God as a higher dimensional being as developed in this chapter? Explain.

Chapter 1.4
Fractals

Fractal mathematics has been gaining popularity over the past thirty years. The subject has gone from being classified as pseudo-mathematics to being studied as mainstream math; people have found applications of fractal theory in computer graphics and cell phone technology. Fractals are both easy and difficult to understand; any child can draw and build fractals, yet fractals' use of infinity produces some challenging, counter-intuitive results.

Warm-up Activity

When you think of the word "infinity," do you usually associate it with the word "big" or "small"? Why? Can you think of ways infinity might be "small"?

Concept Development: Building Fractals

At the most basic level, a fractal is a self-similar object; it can be constructed by using a self-repeating pattern. You will need some paper, pencils, and erasers to build some popular, basic fractals: Sierpinski's Triangle, Cantor Dust, and the Koch Snowflake.

Sierpinski's Triangle. The rule for Sierpinski's Triangle is simply to fill in the middle triangle.

Begin with a triangle. Find and connect the midpoints of each side. Then, shade in the middle triangle.

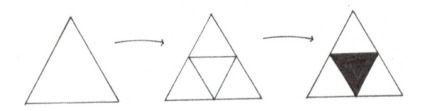

You now have one shaded triangle and three unshaded triangles. Repeat the process for each of the unshaded triangles.

You have now created a second *iteration* of Sierpinski's Triangle. Repeat this process again with the nine unshaded triangles.

Repeat this process again with the 27 unshaded triangles. Continue to repeat this process. You will create Sierpinski's Triangle once you have done this process an infinite number of times.

What are some characteristics of Sierpinski's triangle? How much of the triangle will be shaded? How much of the triangle will be unshaded? We will pursue these questions a little later.

Sierpinski's triangle is a fractal: it follows a repeating, iterated rule that produces self-similarity. (1) Each step of the triangle uses the previous result. (2) The pattern is repeated. (3) If you zoom in on any triangle that was once unshaded, you will see all of Sierpinski's triangle again.

Cantor Dust. The rule for creating the fractal Cantor Dust is also very simple. Whenever you see a segment, erase the middle third.

Each time we repeat the process of erasing the middle third, we are completing another iteration of the fractal. Above you see the 0th, 1st, and 2nd iterations of Cantor Dust. Draw Cantor Dust up to the fourth iteration. At what point does this fractal become difficult to make?

Suppose the first segment is 9 inches long. How much of the segment is left at the first iteration? We erased 3 inches, so 6 inches remain. Cantor Dust is complete when we have gone through this erasing process an infinite number of times. How much of the original segment remains in the complete Cantor Dust?

What makes Cantor Dust a fractal? Again, it follows a repeating, iterated rule that produces self-similarity. (1) Each step of the Dust uses the previous result. (2) The pattern is repeated. (3) If you zoom in on any segment that was once whole you will see all of Cantor Dust again.

Koch Snowflake. The rule for creating the Koch Snowflake fractal is a bit more complicated. On any segment, erase the middle third and replace it with a "peak" whose two slopes are each the same length as the piece removed.

Each of the segments on the triangle is replaced by four shorter segments with a "peak."

Repeat this process for EVERY new segment:

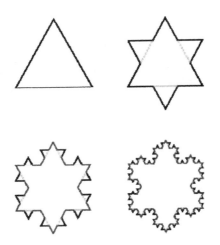

"KochFlake". Licensed under Creative Commons Attribution-Share Alike 3.0 via Wikimedia Commons - http://commons.wikimedia.org/wiki/File:KochFlake.svg#mediaviewer/File:KochFlake.svg

Now add two more iterations to your Koch Snowflake.

Just like the first two fractals we made, the Koch Snowflake follows a repeating, iterated rule that produces self-similarity. However, it is a little different than Sierpinski's

Triangle and Cantor Dust. When the Koch Snowflake is complete, you can zoom in on any portion of the fractal and you will see the same infinitely bumpy snowflake structure.

http://commons.wikimedia.org/wiki/File:Flocke.PNG

The History of Fractals

If you conduct a simple online search for fractals you will find two-dimensional and three-dimensional geometric fractals. You will also find many colorful complex shapes related to the Mandelbrot Set.

Benoît Mandelbrot (1924 – 2010) was a pioneer in fractal mathematics. Mendelbrot's work in fractals did not get much attention from experts in the mid and late 1900's. This was mostly because classic geometry favors smoothness: curves are smooth, polygons are smooth, etc. Furthermore, the basis of Calculus and the analysis of the real number line demands a smooth number line with no "holes." But look at the fractals you created; "smooth" and "continuous" are not words we would use to describe them. The Koch Snowflake is infinitely bumpy. Cantor Dust and the Sierpinski's Triangle are full of holes.

Consequently, until very recently, fractals were considered grotesque and unnatural.

When Mandelbrot published *The Fractal Geometry of Nature* in 1982, he was suddenly thrust into public and academic fame. In this book argued that far from being unnatural, fractals describe nature better than smooth, classical geometry. Because of Mandelbrot's work, since the 1980's fractals have found consistent applications in descriptions of natural phenomena such as plant growth, sea coasts, and mountain ranges.

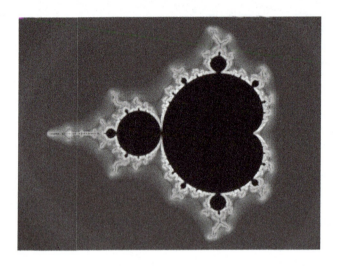

The Mandelbrot Set

http://commons.wikimedia.org/wiki/File:Mandelbrot_set_rainbow_colors.png

The Mandelbrot Set

The image above is a geometric image of the fractal called the Mandelbrot Set. However, the Mandelbrot Set is defined algebraically, not geometrically. It algebraically produces a repeating, iterated rule that produces self-similarity. We must consider two concepts to develop an understanding of what we see here: *complex numbers* and *recursion*.

Mathematical Explorations | 55

Benoît Mandelbrot presenting the Mandelbrot Set
http://commons.wikimedia.org/wiki/File:Mandelbrot_p1130876.jpg

The *complex plane* is the canvas on which the Mandelbrot Set is painted and *complex numbers* are the medium used to create it. You may or may not recall imaginary numbers from your algebra courses. Imaginary numbers emerge as solutions to equations. Consider these examples. If we solve the following equations, we find solutions to each equation.

The solutions to $x^2 = 9$ are $\sqrt{9} = 3$ and $-\sqrt{9} = -3$
The solutions to $x^2 = 2$ are $\sqrt{2}$ and $-\sqrt{2}$
The solutions to $x^2 = -1$ are $\sqrt{-1}$ and $-\sqrt{-1}$

But $\sqrt{-1}$ is not a *real* number; it cannot be plotted on the real number line. While that remains true, $\sqrt{-1}$ *is* one of two solutions to this algebraic equation. Mathematicians therefore named $\sqrt{-1}$ an "imaginary" because it is not on the real number line and gave it the designation of "i."

$$\sqrt{-1} = i$$

These mathematicians also thought of a clever way to locate or plot a number like this using the two-dimensional plane instead of just the real number line.

This is where we use what are called *complex numbers*. A complex number can be of the form $a + bi$ where the number has a real part "a" and an imaginary part "bi." For example, the number $3 + 4i$ has a real value of 3 and an imaginary part of $4i$. It is equivalent to $3 + 4i = 3 + 4\sqrt{-1} = 3 + \sqrt{-16}$. This is also one of the solutions to the quadratic equation: $x^2 - 6x + 25 = 0$. But what gets tricky is plotting the number $3 + 4i$; how can we plot a complex number?

Consider a two-dimensional Cartesian plane, which you have seen before in mathematics class.

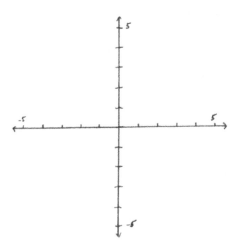

We use this same Cartesian plane to plot imaginary numbers, but we will treat the horizontal axis as the "real axis" and the vertical axis as the "imaginary axis." For example, where the point (3,4) would normally be plotted, the point $3 + 4i$ will be plotted.

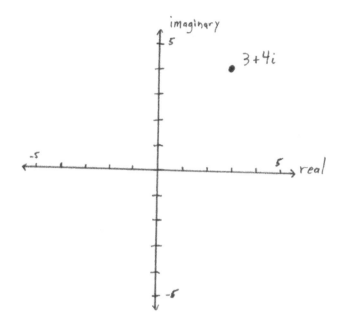

As you may have guessed by now, the Mandelbrot Set uses complex numbers. The other feature we need to examine to understand the Mandelbrot set is *recursion*.

We have briefly talked about recursion before in our consideration of the Fibonacci sequence. Most of our high school math classes used explicitly defined functions instead of recursively defined functions. If we were to find the product of 2 multiplied ten times, we could use the explicit formula:

$$T_n = 2^n \text{ and find } T_{10} = 2^{10} = 1024$$

In other words, we can simply plug "10" into the formula to find 2 multiplied ten times. However, we could also find the value 1024 by using a **recursive** formula:

$$S_n = \begin{cases} 2 \times S_{n-1} \\ S_1 = 2 \end{cases}$$

This recursive process says that we start with "2" ($S_1 = 2$) and then multiply each result over-and-over again by 2.
$S_1 = 2$

$$S_2 = 2 \times S_{2-1} = 2 \times S_1 = 2 \times 2 = 4$$
$$S_3 = 2 \times S_{3-1} = 2 \times S_2 = 2 \times 4 = 8$$
$$S_4 = 2 \times S_{4-1} = 2 \times S_3 = 2 \times 8 = 16$$
$$\vdots$$
$$S_{10} = 2 \times S_{10-1} = 2 \times S_9 = 2 \times 512 = 1024$$

Just like the fractals you constructed earlier, each step in a recursion uses the previous step to produce the next one. The Mandelbrot Set fractal uses a similar recursive formula; it uses a formula that repeats itself over-and-over again. The formula is

$$z_n = (z_{n-1})^2 + c$$

This formula takes the previous result, squares it, and then adds "c." This "c" is also the starting number.

For instance, if $c = 2$, then

$$z_1 = 2$$
$$z_2 = (2)^2 + 2 = 6$$
$$z_3 = (6)^2 + 2 = 38$$
$$z_4 = (38)^2 + 2 = 1446$$

Et cetera

The Mandelbrot Set takes every point in the complex plane and inserts it into this formula. For the sake of simplicity, we will not discuss what it means to multiply imaginary numbers, but we take time to insert real numbers into this formula.

Notice that when we begin with "2," the numbers in the recursion get larger and larger. When beginning with "2," we say that the results "diverge" or go off to infinity. Is this true for all numbers? Consider $c = 0$. If we cycle 0 into the formula, the result is always "0." So, not every number results in a divergent recursion. But "0" is a trivial case; it's almost always peculiar. Let's now consider $c = -\frac{1}{2}$.

$$z_1 = -\frac{1}{2}$$

$$z_2 = \left(-\frac{1}{2}\right)^2 + \left(-\frac{1}{2}\right) = \left(-\frac{1}{4}\right)$$

$$z_3 = \left(-\frac{1}{4}\right)^2 + \left(-\frac{1}{2}\right) = \left(-\frac{7}{16}\right)$$

$$z_4 = \left(-\frac{7}{16}\right)^2 + \left(-\frac{1}{2}\right) = \left(-\frac{79}{256}\right)$$

If we begin with $c = -\frac{1}{2}$, the result does not diverge to plus or minus infinity. Instead, the sequence remains somewhere between $-\frac{1}{2}$ and 0, or $-\frac{1}{2} < x < 0$.

Now you see that there are two types of numbers on the complex plane: those that diverge under this formula and those that do not. This is the basis for the geometric construction of the Mandelbrot Set. If we put a complex number into the formula and it does not diverge, then it is plotted as a black point. If, when put into the formula, a complex number diverges, then it is plotted as a colored point. A number's color is chosen based on how quickly the number diverges.

You now understand what it means that the Mandelbrot Set follows an algebraic, repeating, iterated rule that produces geometric self-similarity. The Mandelbrot Set is a fractal.

Connections: Computer Graphics and Mobile Technology

You may begin to think that fractals are "mathematically cool" but have no real world applications. Actually, mathematicians and scientists continue to discover that some phenomena are more easily modeled by fractals than classic geometry. Consequently, fractals are emerging as important in modern technologies. Two examples of these many applications are mobile technologies and computer graphics.

Mobile Technologies. Although we usually cannot see them from the outside, mobile devices all require antennas to work. Fractals have proven to be effective antennas for two main reasons. First, due to the way a fractal may be folded in on itself, you can twist a long, thin wire into a very small area. As the following exercises will show you using the Koch snowflake, the snowflake can be confined to a very small area, but the length or perimeter of the snowflake is extremely long. The second reason fractals work so well as antennas is their ability to receive a very wide range of signals. As cell phones evolved into smart phones, mathematicians and engineers designed newer technologies to receive phone signals, data signals, GPS information, and other forms of EM waves. Unfortunately, the technology makers found that each kind of signal is picked up more clearly by a different kind of antenna. This meant mobile devices would need multiple antennas. However, fractals saved the day because they allow a single antenna to pick up all of these wave forms.

Computer Graphics. Classic animated cartoons involved drawing thousands of frames to animate the most simple of landscapes. However, as you probably know, computer graphics (computer-generated imagery – CGI) have made the ability to twist and turn the animation simply a matter of a computer algorithm. This sort of motion or feeling that you are flying, soaring, or dropping while watching a movie requires no fractals. However, computer graphic designers have found that when they want to model natural landscapes such as mountains, snowdrifts, flowing water, and large forests, fractals are actually the path to simplicity.

Mandelbrot himself found and demonstrated that , by using a fractal process in our technology, we could create amazing, life-like landscapes in seconds. For example, the landscape constructed below is made by using the same rule on every triangle.

Mathematical Explorations | 61

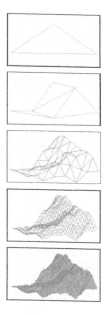

Many of the mountains, lava flows, rivers, and even full planets found in Hollywood films today are produced using fractals.

Something to Think About

Infinity Small. We began this section with some questions: When you think of the word "infinity" do you usually associate it with the word "big" or "small"? Why? Can you think of ways infinity might be "small"?How has this study of fractals impacted your answers to these questions? A fractal is an important gateway into thinking of infinity related to small things. The Sierpinski triangle, the Koch snowflake, and Cantor Dust are not really complete until the patterns are repeated infinitely small. If we were to zoom in on portions of any of these three fractals, we would see the same fractal over and over and over again.

The nature of these fractals leads to some interesting philosophical questions. Does the Sierpinski triangle or the Koch snowflake or Cantor dust actually "exist"? Some people would answer "no" since we cannot truly complete the infinite number of steps to make the infinitesimally small pieces of these three fractals. After all, in physical reality, we can zoom in to the cellular level, zoom in more to the molecular level, zoom in more to the atomic level, and zoom in even more to the subatomic world of up quarks, down quarks, neutrinos, and the whole particle zoo, but even there we are not approaching the "smallness" found in these fractals. Even the theoretically shortest measurable length—the Planck length, named after its developer Max Planck—which measures about 1.6×10^{-35} meters, would contain the whole of Cantor dust.

And yet other people will answer the same existence question, "Yes, these fractals exist in the mathematical sense. I can accurately, mathematically describe them and envision them. In the same way a circle exists, a fractal exists. After all, there are no actually perfect circles in the natural world. Every single one of them is very bumpy at the atomic and subatomic level. Yet the construct of the circle exists in my mind and your mind and can be accurately, mathematically described."

These two answers leave us with more questions. In fact, every time we deal with infinity in mathematics, whether infinity large or infinity small, we wrestle with certain difficult philosophical questions about the nature of reality.

Order. Many have described God as a God of order. What is the nature of this "order"? Is it "order" even if we do not recognize it? Fractals consist of a different kind of order. Does the natural order include fractals? Are fractals a creation of man or a creation of God?

Fractal Landscape – CGI
http://commons.wikimedia.org/wiki/File:FractalLandscape.jpg

Covering the Reading

1. Construct the 4th iteration of Cantor Dust, Sierpinski's Triangle, and the Koch Snowflake.

2. Plot $5 + 8i$ on the complex plane.

3. Would "3" be black or colored in the Mandelbrot Set? Would "1/2" be black or colored in the Mandelbrot Set?

4. Among Cantor Dust, Sierpinski's Triangle, and the Koch Snowflake, which do you believe would be easiest to manufacture as a mobile device antenna? Explain why you chose the one you did.

Problems

5. Research and construct the "Cross Fractal" and "Sierpinski Carpet" fractals.

6. Develop your own fractal rule for a segment or triangle. Show your rule through 3 iterations.

7. Consider Sierpinski's Triangle. Suppose the area of the 0th iteration is 1. On the 1st iteration, $\frac{1}{4}$ of the triangle is shaded. On the 2nd iteration, $\frac{1}{4}$ of the remaining $\frac{3}{4}$ is shaded. At this interation, a total of $\frac{1}{4} + \left(\frac{1}{4}\right)\left(\frac{3}{4}\right) = \frac{7}{16}$ of the triangle is shaded. Continue the pattern to investigate this question: How much of Sierpinski's triangle will be shaded on the 3rd iteration? 4th iteration? Nth iteration? When it is complete?

8. Consider the Koch Snowflake. Suppose each segment on the original triangle in the 0th iteration has a length of 1. At the 0th iteration, the snowflake has a perimeter of 3. At the 1st iteration, $\frac{1}{3}$ of each side is removed and replaced by 2 pieces, each piece equal to the length removed. So at the 1st iteration, the length of each side is $1 - \frac{1}{3} + 2\left(\frac{1}{3}\right) = \frac{4}{3}$ and the perimeter of the snowflake is $\frac{4}{3} \times 3 = 4$. Repeat this process to find the perimeter for the 2nd iteration, 3rd iteration, and 4th iteration. The area of the snowflake is finite (we can fit it in a large square). What do you project the perimeter to be?

9. Do the fractals described in this section "exist"? Explain.

10. Many have described God as a God of order. Fractals consist of a different kind of order. Does the natural order include fractals? Are fractals a creation of man or a creation of God?

Chapter 1.5
Non-Euclidean Geometry

In order to answer the question "What is a non-Euclidean geometry?", it may be helpful to review what Euclidean geometry is. Euclidean geometry is the geometry studied in middle and high schools throughout the world. It is filled with the shapes, properties, and theorems that are familiar to you. For 2000 years, it was the only geometry. Now, Euclidian is one geometry among many others. In this chapter, we will not only study geometry - singular, but several different geometries.

Warm-up Activity

To help you refresh your geometry, take this little Euclidean geometry quiz.

1. What is the name of a triangle with two sides of the same length?
2. What is the difference between a trapezoid and a parallelogram?
3. What is the formula for the area of a circle?
4. What shapes make up the surface of a cylinder?
5. What is the measure of an angle that is the compliment of a 30 degree angle?
6. What is the measure of an angle that is the supplement of a 30 degree angle?
7. In any right triangle, what is the relationship among the sides? (Hint: this is the Pythagorean Theorem)
8. What is the sum of the interior angles of any triangle?

Euclidean Geometry

Hopefully you did well on the Euclidean geometry quiz. If not, you could easily find the answers in any geometry text. The properties and theorems related to your answers are the results in Euclidean geometry. However, at its heart, Euclidean geometry is something much deeper. Although it was developed 2000 years ago, it models a structure that all mathematics follows: an *axiomatic system*. An axiomatic system is one that begins with some assumptions and then uses logic to produce results. Let's contrast an axiomatic system with an inductive approach to better understand what Euclidean geometry has to offer.

What axiomatic structure is NOT: Suppose I asked a group of people to do the following activity:

1. Draw a triangle.
2. Measure the interior angles with a protractor.
3. Add up the three angles.

I have done this activity many times in class. Here are some typical results:

```
181   178   180   180   179
  181   180   182   179   180
181   180   179   180   179
  180   181   180   180   182
```

Now, I ask you to make a *conjecture*. A conjecture is a kind of educated guess based on data. Complete the following conjecture (You probably know the answer from the warm-up quiz):

Conjecture: The sum of the interior angles of any triangle is _____ *degrees.*

During this activity, you have studied geometry *inductively*. We could study all of geometry inductively. One of my favorite high school geometry texts is Michael Serra's *Discovering Geometry* because he leads the reader through an inductive study of the entire high school geometry curriculum. However, geometry itself—that is Euclidean geometry—is *not* inductively constructed. Induction is a fun way to learn about the properties of geometry, but it does not represent *axiomatic structure* of geometry.

What axiomatic structure IS. In Euclidean geometry, we come to the same conclusion: the sum of the interior angles of any triangle is 180 degrees. However, we do not approach it through inductive, observational, scientific-like reasoning. Instead, we begin with a set of axioms or postulates (statements accepted as true without a mathematical proof) and then build all of the logical consequences of those axioms.

Euclid's *The Elements* are a set of 13 books written by the mathematician Euclid around 300 B.C. in Alexandria, Egypt. In his books, Euclid systematized all known geometry up to that time. His work was so significant and long-lasting that anyone who studied geometry up through about 1900, including President Abraham Lincoln, used Euclid's *Elements* as their geometry textbook. Euclid *was* geometry, and geometry was the basis for Sir Isaac Newton's work in physics. Hence, Euclidean geometry was the foundation for understanding our universe.

Euclid began with 5 Axioms or Postulates that he considered so obvious that no one could ever question them. He also identified some Common Notions and provided starting definitions. All of geometry was *deductively* constructed from Euclid's foundations. In other words, every theorem was a *logical* consequence of the axioms and other theorems. There could be no evidence that could be contrary to the conclusions because the conclusions were not based on evidence. The conclusions were based on reason.

Starting points. What were the starting points that Euclid chose that required no evidence? Here are the 5 axioms stated in modern language:

1. A line segment can connect any two points.
2. A line segment can be extended indefinitely in a straight line.
3. A line segment can be used to draw a circle with the radius the length of the line segment.
4. All right angles are congruent.
5. Given a line and a point *not* on that line, exactly one line can be drawn through the point that is parallel to the line.

These axioms should seem to be a very safe place to begin. Basically, they say you can draw lines, circles, right angles, and parallel lines. Euclid considered these to be so basic—so fundamentally true—that any theorems that logically resulted from these would also be true.

However, you may have noticed earlier that he also added some definitions and common notions. These definitions allow us to basically refer to the same ideas when we communicate, but the common notions are more like axioms. Here is a sample of Euclid's Common Notions in modern notation. I think you will easily agree with them:

Example 1: if a=b then b=a
Example 2: if a=b and b=c then a=c
Example 3: if a=b then a+c=b+c

From these axioms, common notions, and definitions, Euclid employed logic to produce *theorems*. Theorems are very different than theories. A theory is a generalization that fits all of the evidence. If new evidence arises, a theory may be adjusted to fit the evidence. This is not the case with theorems. A theorem is a property that *must* be true as a logical consequence of the axioms. In other words, theorems are absolutely true, as long as the logical argument used to produce them is *valid* and the assumptions or axioms on which the theorem is based are *sound*.

	Axiomatic Structure			
Consequences	Theorem 5			Logic
	Theorem 4			
	Theorem 3			
	Theorem 2			
	Theorem 1			
Assumptions	Definitions	Axioms	Common Notions	

Let us return to our conjecture/theorem: The sum of the interior angles of a triangle equals 180 degrees (Triangle Sum Theorem). We first found the sum inductively by taking repeated, error-prone measurements. Deductive, this theorem is a logical consequence of axioms, common notions, and other theorems. Through a series of logical arguments, Euclid proved this theorem necessary for our current argument:

Theorem (Alternate Interior Angles Theorem) Given two parallel lines crossed by a transversal, alternate interior angles are congruent.

In picture form, the theorem looks like this:

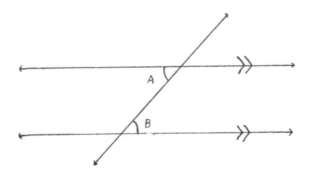

Angle A and angle B are on alternate sides of the *transversal* (the line that cuts across the parallel lines) and on the interior of the parallel lines. Hence, they are alternate interior angles. This theorem, proven through many other logical steps, says that alternate interior angles are congruent—they have the same measure.

What does this theorem have to do with the Triangle Sum Theorem? It is one of the theorems used in the deductive argument for the triangle sum theorem along with some other common notions and axioms.

Here is a Euclidean, logical argument for the Triangle Sum Theorem. First we begin with a triangle.

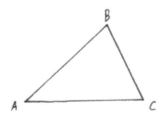

We want to prove that $m < BAC + m < ABC + m < BCA = 180°$. Or simply, angle A + angle B + angle C equals 180. Next, using axiom 5 (Given a line and a point *not* on that line, exactly one line can be draw through the point that is parallel to the line), we can draw a line parallel to line AC through point B. This makes line AC parallel to line EF with two transversals: line AB and line BC.

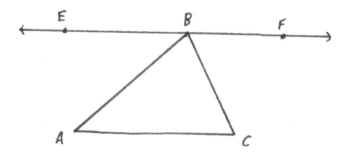

We will now invoke the Alternate Interior Angle Theorem twice. $m < BAC = m < EBA$ and $m < ACB = m < CBF$. You can see these congruencies in this illustration.

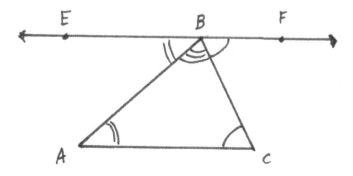

The rest of the proof is fairly straight forward. I will present it first in everyday language, and then formally. The angles in the triangle are the single, double, and triple hash mark angles. Also, on one side of the straight line EF, we see the single, double, and triple hash mark angles. Since one side of a straight line is 180 degrees, the triangle which is made up of the same three angles is also 180 degrees. Now, here is the same proof written more formally:

We are trying to prove:
$m < BAC + m < ABC + m < ACB = 180°$
Since EF is a straight line:
$m < EBA + m < ABC + m < CBF = 180°$
We know by the AIA Theorem:
$m < BAC = m < EBA$ and $m < ACB = m < CBF$.
By substitution (a common notion):
$m < BAC + m < ABC + m < ACB = 180°$

Hence, the sum of the interior angles of any triangle is 180 degrees.

It is worth taking time to correct a possible misconception at this point. Most people use the words "evidence" and "proof" interchangeably. However, in the field of mathematics, the two words are dramatically different. If you are still using these words interchangeably, you may want to reread portions of this chapter after reading this paragraph.

If you make an observation of an individual case, then you have evidence that something is true. Many similar observations produce a preponderance of evidence that something is true. However, evidence *could* be presented to the contrary that could upset or modify the truth of the claim. Mathematical proof is different than evidence.

A mathematical proof is a logical argument that something must be true given that something else is true. So mathematical structure uses proof to establish the truth of something, based on something else, based on something else, based on something else, and so on. At some point, the path has to end and be based on something that is accepted true without proof. Euclid's Postulates are statements accepted as true without proof and serve as the starting point for all other Euclidean geometric proofs. This approach is axiomatic: begin with axioms, statements accepted true without proof, and move logically to consequences that must be true.

Euclid's axioms are accepted true without proof, but there is plenty of evidence that they are true. In fact, Euclid felt that the evidence for his five postulates was so strong that they were a safe place on which to build the entirety of geometry. The axioms serve as a foundation for the whole building; the truth of every theorem rests on the truth of the starting assumptions. A non-example will help to clarify the precariousness of an axiomatic structure. Observe how the strength of the structure is totally based on the strength of the foundation:

Axiom: $2 + 2 = 5$

Theorem: Magical Unicorns are Real

Proof:

1. $2 + 2 = 5$ (Axiom)
2. $4 = 5$ (Property of addition)
3. $0 = 1$ (Add -4 to both sides of the equation- Addition Property of Equality)
4. A horse with 0 horns has 1 horn (since $0 = 1$).
5. $0 = 1,000,000$ (multiply both sides of equation 3 by 1,000,000)
6. None of something is the same as a lot of something (Consequence of equation 5).
7. An animal with no magic has a lot of magic (Consequence of line 6).
8. There exist magical unicorns (Consequence of line 4 and 7)

Historical Interlude

During the 600 years between the mathematical renaissance in the West and the late 1800's, many mathematicians undertook a quest to "prove the 5th." They realized, as you might have, that Euclid's first four postulates are sleek and simple while the 5th postulate is cumbersome and clumsy. The version of the 5th presented in this chapter, attributed to the mathematician Playfair, is actually already a sleeker version of Euclid's clumsier statement. For the sake of clarity, I have restated the five postulates/axioms here in as basic a form as possible:

1. Two points can produce a line segment.
2. A line segment can produce a line.
3. A line segment can produce a circle.
4. All right angles are congruent.
5. Given a line and a point *not* on that line, exactly one line can be drawn through the point that is parallel to the line.

Even in its simplest form, Axiom 5, called Euclid's 5th Postulate or The Parallel Postulate in most literature, looks more like a theorem than an axiom. Consequently, many mathematicians undertook the challenge of trying to prove the 5th from the other four. Why? Two reasons. The first is for beauty's sake; the 5th postulate is too ugly to be an axiom. The second reason is structural reason. Remember that an axiomatic system is a totally secure system that establishes certain truths to the extent that its axioms or starting assumptions are certain. Euclid claims to begin with only five assumptions accepted true without mathematical proof. If mathematicians could prove that the 5th is a logical consequence of the other four—that is that the 5th is a theorem—then all of geometry can be constructed with only *four* assumptions instead of *five*. From a structural standpoint, that accomplishment would make something already quite safe and certain even more certain. Euclid's mathematics would then border on absolute perfection.

Now, remember earlier that we talked about geometries, as in more-than-one? Carl Friedrich Gauss (1777-1855) was the pioneer in non-Euclidean geometry. His work began a revolution in mathematics that changed our perception of certainty in the subject. As with most ground-breaking activities that challenge everyone's preconceived notions, Gauss' developments were radical and unwelcome at the time. Philosopher Immanuel Kant argued in 1781 that nothing could be more certain: "Thus, moreover, the principles of geometry—for example, that 'in a triangle, two sides together are greater than the third,' are never deduced from general conceptions of line and triangle, but from intuition, and this a priori, with apodeictic certainty." By the way, "apodeictic" means "clearly beyond dispute." Given its reliability, its relationship to Newtonian mechanics, its use of cool logic, and the supposed obviousness of its results, Euclid's *Elements* was as sacrosanct as the Holy Scriptures.

But Gauss was one of the most mentally powerful and insightful mathematicians of all time. He pressed forward in many areas of mathematics including establishing probability's basis for statistics through the normal curve, gaining an understanding of the distribution of prime numbers, and solidifying number theory as a discipline. He frequently took other mathematicians proofs and showed that he could do them in a more elegant way; in a sense, he embodied the phrase "everything you can do, I can do better" In the field of mathematics. That seems a bit arrogant, but he truly stood head and shoulders above his peers. Nonetheless, when he developed non-Euclidean geometry, he hesitated to share his results with the larger mathematical community, fearing the uproar of those "Boeotians" (That's an insult to the other mathematicians—Boeotians were people from a region of Greece thought to be dull-witted).

What did Gauss do with geometry? For a time Gauss devoted himself to the problem of proving Euclid's 5th from the other four postulates. Many mathematicians had given it a try, and it was Gauss' turn. It turned out that even the brilliant, usually successful Gauss was foiled as well. Yet Gauss' attempt led him in a different direction than all the other mathematicians: when Gauss could not prove the Parallel Postulate from the other four, he began to question the *truth* of the postulate itself. I do not know if it was out of arrogance or if he just had an amazing insight, but he was the first to seriously question the Parallel Postulate's veracity. Even Gauss' teacher at one time said that only a fool would ever question the Euclid's 5th. And then Gauss did.

In letters to his brilliant student the Hungarian Janos Bolyai, Gauss explained that his work on Euclid's Parallel Postulate led him not only to fail to prove the 5th, but that this failure made him question the truth of the Parallel Postulate. Furthermore, Gauss believed he had established a whole new system of geometry in which the Parallel Postulate became something completely different. In other words, Gauss kept Euclid's axioms one, two, three, and four and replaced Euclid's 5th Postulate with a different version of the postulate. Using this new set of postulates he was able to establish a seemingly bizarre but internally consistent geometry that was just as strong as Euclid's. Bolyai built on Gauss' work and went on to "dot the i's and cross the t's" in one type of non-Euclidean geometry.

http://en.wikipedia.org/wiki/File:Janos_Bolyai_memorial_plaque.jpg

Non-Euclidean Geometry

Instead of directly attacking or changing Euclid's 5th Postulate, Gauss looked at one of the consequences of the Parallel Postulate. Remember the proof that the sum of the interior angles of a triangle equals 180 degrees (Triangle Sum Theorem)? In that proof, Euclid's 5th Postulate is used to construct a parallel line. Then, using that parallel line, we invoked the Alternate Interior Angle Theorem, which in turn is used to prove that the angles add up to 180 degrees. Since they are so often used together like this, Euclid's Parallel Postulate and the Triangle Sum Theorem are intimately linked. But what would happen if you changed the Triangle Sum Theorem?

In Euclid, the Triangle Sum Theorem states:
The sum of the interior angles of a triangle equals 180 degrees.

Gauss investigated what would happen if he changed the theorem to say:
The sum of the interior angles of a triangle *is less than* 180 degrees.

Your intuitions may say that what Gauss did was crazy and that the change he made is totally inconsistent with geometry. You would be in part correct. The theorem cannot simply be changed by fiat without destroying the whole internal consistency of Euclidean geometry. Actually, most people conclude that Gauss' move would result in total nonsense.

However, Gauss pursued his change, and it yielded many consequences related the Triangle Sum Theorem. For example, in Euclidian geometry, since the sum of the interior angles of a triangle is 180 degrees, the sum of the interior angles of a quadrilateral is 360 degrees. If Gauss changes the Triangle Sum Theorem, he also has to change the theorem related to quadrilateral. Essentially, he would have to change a whole host of theorems. At some point, mathematicians expected him to begin to develop statements that were logically *inconsistent*—that is, two statements that could not both simultaneously be true. (Here is an example of two inconsistent statements: "A tangent is perpendicular to the radius of a circle" and "A tangent is not perpendicular to a diameter of a circle." Since a diameter is made up of two radii, the two statements are incompatible.) However, as Gauss worked, he changed theorem after theorem to adjust to his experimental "less than 180 degrees" and never found inconsistent statements. He ended up morphing Euclid's 5th Postulate as demonstrated below:

Euclid's 5th Postulate (Playfair's version):
**Given a line and a point *not* on that line,
<u>exactly one line</u> can be drawn through the point that is
parallel to the line.**

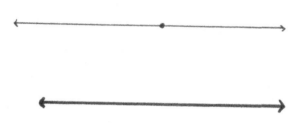

The 5th Postulate consistent with Gauss' "less than 180 degrees":
**Given a line and a point *not* on that line,
<u>infinitely many lines</u> can be drawn through the point
that are parallel to the line.**

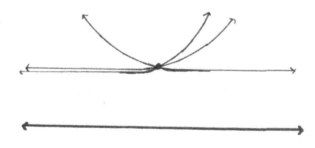

This new geometry, later known as Hyperbolic Geometry, is just as consistent—just as mathematically solid—as Euclidean Geometry. Euclid begins with five postulates: four postulates plus the parallel postulate. Hyperbolic Geometry begins with five postulates: Euclid 1-4 and a modified 5th postulate. The modification has to be specific, but the result is another whole system that is just like Euclidean Geometry, but entirely different!

You may be thinking that the "lines" under Hyperbolic Geometry are not lines. You are both right and wrong. They are not the lines that we experience in Euclidean Geometry, but they are lines in Hyperbolic Geometry. This conclusion is not intuitive because they do not look like lines as you are used to experiencing them. However, several natural models allow us to see that they are actually lines. What has changed is the nature of the surface on which we are constructing geometric argument.

In order to understand non-Euclidean geometries more intuitively, let's explore a second non-Euclidean geometry: Elliptical or Spherical Geometry. In Spherical Geometry, Euclid's 5th postulate is also adjusted, but with a different modification:

> Given a line and a point *not* on that line,
> <u>zero lines</u> can be drawn through the point
> that are parallel to the line.

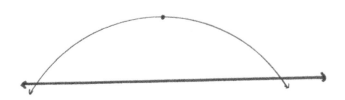

When we use Spherical Geometry in relation to the triangle, the sum of the interior angles of a triangle is *greater than* 180 degrees. Again, this does not seem to be consistent because we want to insist that there are no a straight lines; however, this familiar model allows us to conceptualize Spherical Geometry.

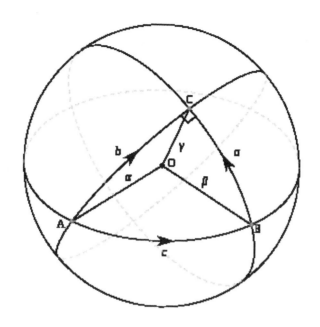

Imagine yourself on the surface of a sphere: the planet Earth. What does it mean to walk in a "straight line"? In all geometries, point, line, and plane are undefined terms. They exist in whatever context we find them. So, on the surface of a sphere, what does it mean to go in a straight line? If you walk in a straight line on the surface of a sphere in *any* direction, you come back to where you started. A line on a sphere is what we may call a "great circle." The equator is a familiar example.

Now that you can visualize a Spherical geometry, let's rethink our changes to Euclid's 5th Postulate. Any pair of Spherical Lines that you draw on the surface of the sphere will intersect. Hence, given a line and a point not on a line, every line intersects with every other line! There are no parallel lines. This illustration allows us to visualize an easy application of Spherical Geometry.

You may argue that the horizontal "lines of latitude" are parallel lines. However, lines of latitude are not lines on a sphere.

They are actually circles. To walk along a line of latitude you must constantly be turning. You will not be walking "straight."

But how does this relate back to the triangle theorem? How can it be that the sum of the interior angles of a triangle is greater than 180 degrees? Again, take a trip on the sphere. Start at the North Pole. Head south to the equator. Turn left 90 degrees. Travel some miles along the equator. Turn left 90 degrees and head toward the North Pole. When you arrive, you will see the footprints you made when departing and when returning. These two paths will form an angle. For simplicity's sake, let's suppose that angle is 30 degrees. You have traveled in a spherical triangle. The sum of the angles is 90 + 90 + 30 = 210 degrees.

Similarly, hyperbolic geometry operates on a different "surface." The hyperbolic plane is described as saddle-shaped and is more difficult to model than the elliptical plane (the globe). However, on the hyperbolic plane, a given line has multiple lines parallel to it. The angles of a triangle add up to less than 180 degrees. While the shape of space is different and more difficult to model, try visualizing it and making sense of this hyperbolic world and its rules.

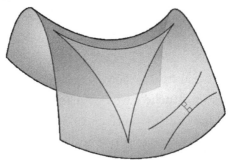

Something to Think About

Non-Euclidean Geometries: Real or Not? True or Not? Even though they were eventually accepted as mathematically sound, Hyperbolic Geometry as developed by Gauss, Bolyai, and Lobachevsky and Elliptical Geometry as developed by Riemann and Schlafli were first seen as bizarre mathematical oddities. Mathematicians distinguished only between real-Euclidean geometry and everything else (yes, even more geometries arose). Euclidean geometry described the "real world" while Hyperbolic and Elliptical geometries did not. However, this misconception did not survive.

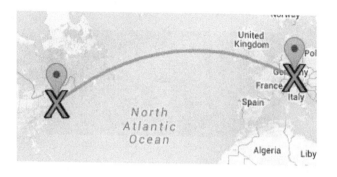

Many of us have experienced Spherical/Elliptical Geometry and been puzzled by it. When I first traveled from New York to Munich, Germany, they still displayed the flight path on screens on the airplane. Now flights can easily be tracked from the internet. It puzzled me that instead of heading straight across the ocean to France, we seemed to be going the "long way" up over Canada, skirting southern Greenland, and coming down over Britain into Germany. Why would we take the long path? It turned out that the northern route was actually the short path: the shortest distance between two points is a straight line. On a sphere, a "straight line" is the great circle of Spherical Geometry. Stretch a string straight from New York to Munich on a globe. Pull it as tight as you can. What is the shortest path? The shortest path is the "long northern route" just as predicted by Spherical Geometry. So now let's ask the question again: which is the real geometry that accurately portrays the world: Euclidean or Spherical?

If Spherical Geometry has real world applications, maybe Hyperbolic Geometry does, too. In the early 1900's, a brilliant scientist named Albert Einstein was searching for a mathematical model that accurately models his Theory of General Relativity. General Relativity describes the shape of the universe. How does gravity bend the space-time? How can something like light travel in a "straight line" and yet also "curve"?

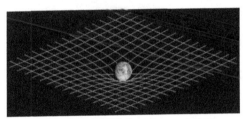

http://en.wikipedia.org/wiki/File:Spacetime_curvature.png

The answer turned out to be quite simple: mathematicians dusted off some old Hyperbolic Geometry texts and Einstein had his model. Light does travel in straight *hyperbolic* lines. Gravity bends space-time in a way that is described most accurately and simply by Hyperbolic Geometry. So the question remains—which is the real geometry that accurately portrays the world: Euclidean? Spherical? Hyperbolic?

Assumptions. The study of non-Euclidean geometries teaches us that beginning with different starting assumptions can result in very different, yet consistent, results. People face the same issues. Frequently, I have found that when I disagree with someone about an issue or even a belief, I at first find it hard to understand how a thoughtful, intelligent person can have such a different view than mine. Learning such as the kind we have done in this chapter has taught me to pursue a path other than immediately assuming one of us is a fool. When I investigate *why* someone comes to the conclusion they do, I often find those conclusions to be as logical as my own. I have seen this trend in minor disagreements, theological tensions, and significant differences of opinion. What is key is that the other person and I have different starting *assumptions*. For instance, I begin believing that there is a God and that he communicated to us through the Bible. If someone starts with different assumptions, we should not be surprised that we disagree significantly; we are proceeding through different systems of thought. This realization makes me more reflective about the basic assumptions that guide my decisions in life. I often ask myself how my assumptions differ from other people's. More importantly, I must evaluate the reliability of my starting assumptions; why are my assumptions, and therefore conclusions, correct?

Covering the Reading

1. In your own words, explain the link between parallel lines and the Triangle Sum Theorem.

2. In your own words, explain what an axiomatic system is.

3. In your own words, explain what it means for a geometry to be "non-Euclidean."

Problems

4. How are Euclidean Geometry and Spherical Geometry the same?

5. How are Euclidean Geometry and Spherical Geometry different?

6. How are Spherical Geometry and Hyperbolic Geometry the same?

7. A triangle is drawn on a sphere. Can you determine the size of the angle **x**? Why or why not?

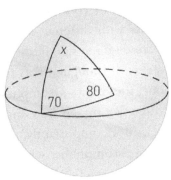

8. What does the sentence "There are different geometries" mean? What is a geometry? How can there be different ones?

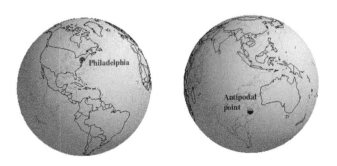

9. The antipodal point from Philadelphia (the point that lies on the opposite side of the world) is plotted below. It is a (wet) place with coordinates 39°59'53" South and 104°51'19" East. An airplane pilot wishes to fly there directly from Philadelphia. He must choose a heading for the direction of his flight path. How many directions are available for a shortest journey? Can two

points on a Euclidean plane have that same number of shortest paths? Explain the difference or similarity of your two answers.

10. In this chapter we looked at 3 different geometries. Which one is "right"? Defend your position.

11. One of the major impacts of studying different geometries is a deeper understanding of what an axiomatic system is and how it works. In this chapter, we saw that a change in starting assumptions can result in significantly different *logical* conclusions. Two parties can come to diametrically opposed positions through very logical means *if* their starting points are different. Provide an example outside of mathematics in which beginning with different starting assumptions can logically lead to very different conclusions.

12. What are the basic assumptions that guide your decisions in life? Each person has different underlying beliefs. What your most foundational assumptions from which you derive your conclusions about moral and ethical issues? How might your assumptions differ from someone else?

Chapter 1.6
The Golden Ratio

The Golden Ratio has many names: the Golden Ratio, the Golden Section, the Golden Rectangle, even the "Sexiest Rectangle." It's easy to find many websites devoted to the topic; it even appears in Disney's *Donald Duck in Mathemagic Land* cartoon. Why the interest? People claim that this ratio is imbedded in ancient and modern architecture, art, the human body, many human creations, and throughout nature. So what is this Golden Ratio?

Warm-Up Activity 1
1. Sketch what you believe to be the "nicest looking" or "most appealing rectangle."
2. Measure the length and width of the rectangle to the nearest millimeter.
3. Divide the larger number (the long side) by the smaller number (the short side).
4. Do you get a number near 1.618? If you did, then you drew a Golden Rectangle.

Warm-Up Activity 2
1. Walk around and find 10 rectangles created by people. These could anything rectangular like picture frames and computer screens.

2. Measure the length and width of each rectangle to the nearest millimeter.
3. Divide the larger number (the long side) by the smaller number (the short side).
4. How many are nearly Golden?

Object	Length	Width	Length divided by Width	Near 1.618?

Warm-Up Activity 3

What is the Golden Ratio? It has very special mathematical properties. Try this.

1. Put a point on the line segment below somewhere other than the exact middle. Your divides the whole segment into two other segments: a short segment and a long segment.

2. Measure each segment to the nearest millimeter:
   Whole:
   Long:
   Short:

3. Divide these two ratios: $\frac{Whole}{Long} = \qquad \frac{Long}{Short} =$

4. Are these two values the same? If not, redo the activity until these two values are the same. Once these ratios are the same, you have found a point that divides the segment into the Golden Ratio.

Concept Development

To gain a fuller mathematical understanding of the Golden Ratio, we are going to examine it from five different perspectives. There are many geometric ways to understand the ratio and one related arithmetic way.

The Divided segment. In the previous activity, you discovered the Golden Ratio on a segment. The Golden Ratio is always a ratio of two values. The ratio has a very particular property: $\frac{Whole}{Long} = \frac{Long}{Short}$. The Golden Ratio is usually expressed using the Greek letter φ or φ (phi). If $\frac{Whole}{Long} = \frac{Long}{Short}$, then $\frac{Whole}{Long} = \frac{Long}{Short} = \varphi$. Numerically, $\varphi = \frac{1+\sqrt{5}}{2} \approx 1.61803398$. Now look at the *reciprocal* of φ; it has a very curious decimal expression: $\frac{1}{\varphi} \approx 0.61803398$. Did you notice the similarity? This is not a coincidence. This similarity is the very nature of this $\frac{Whole}{Long} = \frac{Long}{Short}$ relationship. So whether you divide $\frac{Long}{Short} \approx 1.61803398$ or $\frac{Short}{Long} \approx 0.61803398$, the decimal representation is easy to recognize.

Self-repeating rectangle. A Golden Rectangle is a rectangle whose sides are in the Golden Ratio. How does that shape relate to the $\frac{Whole}{Long} = \frac{Long}{Short}$ relationship if there are only two lengths to measure? An infinite number of Golden Rectangles is built into the explanation. Consider this Golden Rectangle:

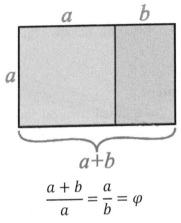

$$\frac{a+b}{a} = \frac{a}{b} = \varphi$$

The vertical segment inside the rectangle is at the Golden Ratio point. The relationship $\frac{a+b}{a} = \frac{a}{b} = \varphi$ parallels the whole:long = long:short relationship. What is really interesting here is that $\frac{a+b}{a} = \varphi$ so the large rectangle is a Golden Rectangle. However, it is also notable that $\frac{a}{b} = \varphi$ indicates that the small light rectangle is also a Golden Rectangle.

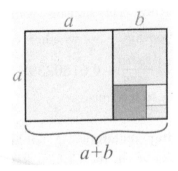

When the Golden Ratio point is found in any Golden Rectangle, the resultant smaller rectangle is also a Golden Rectangle. Therefore, the small light rectangle also contains a Golden Rectangle, which contains a Golden Rectangle, which contains a Golden Rectangle, *etc.*

The Fibonacci Sequence. At first, the Fibonacci Sequence seems an unlikely place to find the Golden Ratio. Remember, the sequence was popularized in a book written in 1202 by Leonardo Fibonacci of Pisa. He used a contrived but natural phenomena of reproducing rabbits to model the recursive, self-referencing nature of this sequence.
Remember, the Fibonacci Sequence is:

$$1, 1, 2, 3, 5, 8, 13, 21, \ldots$$

Can you predict the next few terms? If you can, then you understand what it means to be recursive or self-referencing. In this case, each term is the sum of the previous two terms. More technically:

$$F_n = F_{n-1} + F_{n-2}$$
$$F_1 = 1$$
$$F_2 = 1$$

This is an interesting sequence in and of itself. At first glance, though, it has little to do with the Golden Ratio. However, consider these ratios of consecutive pairs of Fibonacci numbers.

$$1/1 = 1$$
$$2/1 = 2$$
$$3/2 = 1.5$$
$$5/3 = 1.\overline{6}$$
$$8/5 = 1.6$$
$$13/8 = 1.625$$
$$21/13 \approx 1.6154$$
$$34/21 \approx 1.6190$$
$$55/34 \approx 1.6176$$
$$89/55 \approx 1.6182$$

What do you see? Yes! The Fibonacci Sequence relates very clearly to the Golden Ratio: the limit of the rations of Fibonacci Numbers equals φ.

$$\lim_{n \to \infty} \frac{F_{n+1}}{F_n} = \varphi$$

The Fibonacci-produced rectangle. Suppose we built a rectangle using the Fibonacci Sequence. Begin with 2 one-by-one squares next to each other. Put together, these roduce a 2 × 1 rectangle with a $\frac{2}{1}$ ratio. Add a two-by-two square and you produce a 3 × 2 rectangle with a $\frac{3}{2}$ ratio. You can continue to build rectangles using n-by-n squares that follow the Fibonacci Sequence.

Fibonacci	Rectangle	Ratio
1,1		$\frac{2}{1} = 2$
2		$\frac{3}{2} = 1.5$
3		$\frac{5}{3} = 1.\overline{6}$
5		$\frac{8}{5} = 1.6$
8		$\frac{13}{8} = 1.625$

As this pattern continues, the rectangle becomes more Golden with every step.

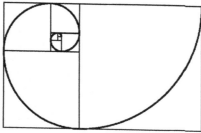

The Golden Spiral. If you smoothly connect one of the new vertices created by the new rectangle in each step above, you trace this spiral, sometimes called the Golden Spiral.

Beyond the Mathematics

The Golden Ratio is inherently mathematically interesting. However, its applications beyond pure mathematics usually attract most people. Many claim that the Golden Ratio is a criteria for beauty imbedded in the very fabric of nature. This beauty criterion has been both intentionally and unintentionally adopted into the art and architecture of our daily lives. Let's examine some of the claims and counter-claims for φ as naturally beautiful and φ in beautiful man-made objects.

The Golden Ratio in natural objects. Many people have recognized the Golden Ratio in natural objects. More specifically, they point to it in at least three forms: the Fibonacci Numbers (1, 1, 2, 3, 5, 8, 13, 21, 34, 55, 89...) appear in natural objects at an alarming rate, the Golden Spiral appears in cross-sections of natural objects, and φ is found in many natural ratios in the human body and other living creatures. An example or two of each type should help you to consider these claims and further investigate them for yourself.

First, the Fibonacci Numbers (1, 1, 2, 3, 5, 8, 13, 21, 34, 55, 89...) are abundant in flowers and plants: Pine Cone (8), Black Eyed Susan (5), Trillium (3), Shasta Daisy (21):

The Golden Spiral seems to appear in a Nautilus Shell cross-section and Sunflower's spiral growth pattern.

The ratios of the human form are said to be Golden. Here are a few samples

Top of head to heel	Divided by:	Top of head to fingertips
Top of head to fingertips	Divided by:	Top of head to navel
Top of head to navel	Divided by:	Top of head to top of inner arm
Top of head to top of inner arm	Divided by:	Top of head to chin
First knuckle length	Divided by:	Second knuckle length
Elbow to fingertips	Divided by:	Elbow to wrist
Center of pupil to center of bottom of teeth	Divided by:	Bottom of teeth to chin

The Golden Ratio in human creations: People claim to find the Golden Ratio in human creations including ancient and modern architecture, artist's masterpieces, and photography.. Consider the examples below.

The Golden Ratio/ Rectangle in architecture: Width compared to height of the Parthenon; ratios in Notre Dame Cathedral

The Golden Ratio/Rectangle in art: *St. Jerome*, the face of the *Mona Lisa*, the location (right and up) within the painting of the central figure's face in *Bathers at Asnières*.

The Golden Ratio in photography: There is a "rule of thirds" in photography clarifying where to locate the central object of your image. This rule of thirds divides the viewing area roughly by the golden ratio:

http://people.oregonstate.edu/~dearingj/the-rules-of-thirds/rule-of-thirds-2/
http://people.oregonstate.edu/~dearingj/the-rules-of-thirds/rule-of-thirds-1/

Counter-claims. Not everyone agrees that φ has a special place in nature, art, and architecture. Some people try to point out that every argument the Golden Ratio people make is flawed. They work to show that many of the approximations are too far off to be φ or that the dimensions have to be mildly morphed to be Golden. Another argument that debunks claims of the Golden Ratio in Nature is that there are as many, if not more, non-examples as there are examples! Plenty of flowers have a non-Fibonacci number of petals. Many spirals are not Golden. Plenty of buildings sport ratios other than the Golden one.

Which side of the argument do you believe?

Something to Think About

We need to process two significant topics in this section. One specifically has to do with the Golden Ratio; the other has more to do with beauty in general.

It is nonsensical to mathematically argue that there is no such thing as phi. What we are really wrestling with are deeper questions of phi's significance or insignificance. Is phi one among many, many ratios found in nature and human design? Or is phi special in nature and human design?

Is phi an objective measure for beauty? Some claim that things that contain the Golden Ratio are more beautiful those that do not. This is why we use the "rule of thirds" in photography and what makes one person more beautiful than another. However, many people say that beauty is "in the eye of the beholder," that is to say that beauty is *subjective*. If phi is a measure for beauty or a way to help determine if something is beautiful, then beauty is at least partially objective. So is phi an objective measure for beauty? Are there any objective measures for beauty? What does God have to do with beauty and our understanding of it? What does sin have to do with it?

Covering the Reading

1. Choose 5 common rectangular objects like index cards, paper, and books. Measure the dimensions and compare them to phi.

2. Conduct some research: Ask 10 people to make a nice rectangle. Measure the dimensions and compare them to phi.

3. Draw a Golden Rectangle using the methods described in the chapter.

Problems

4. Find 3 websites about the Golden Ratio. What claims do they make beyond what is presented in this chapter?

5. Research counter-arguments against the Golden Ratio claims. Summarize the anti-phi arguments.

6. Form and communicate an opinion related to the phi-arguments.

7. Do you believe phi is an objective measure for beauty? Do you believe there are other objective measures of beauty? What does God have to do with beauty? Does He decide what is beautiful and we simply agree, or can we decide? Defend your position.

8. Ecclesiastes 3:11 says that God makes everything beautiful in its time. Is God a measure of beauty? If so, how? Are we the deciders of beauty? Explain. State and support a position on the relationships among beauty, God, and humanity.

Chapter 1.7
Proof and Beauty

A proof is a convincing mathematical argument. In this chapter, we are going to look at proofs of the Pythagorean Theorem. In other words, we will show that the theorem is true for *every* right triangle. However, before we tackle that task, we need to consider what it takes to *prove* something is true in mathematics.

Warm-up Activity

Consider the following two conjectures (a conjecture is something we believe is probably true but want to prove with certainty):

1. If we add two odd numbers, we get an even number.

2. If we multiply two odd numbers, we get an odd number.

It is not difficult to find examples of these two conjectures. But how can we prove that these properties will be true for *all* odd numbers? How can we express *any* odd number, not just a specific odd number? Try to develop an argument that these conjectures are true—turn these conjectures into theorems by proving they are always true.

Concept Development

The best know theorem of all time is probably the Pythagorean Theorem. Even those who remember nothing else about Euclidean geometry remember something about the Pythagorean Theorem.

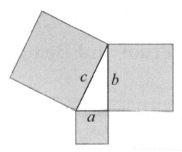

Here is a refresher for you: "the square of the hypotenuse of a right-angled triangle equals the sum of the squares of the other two sides." In modern algebraic notation, the theorem states:

In a right triangle, with side lengths a and b and hypotenuse length c,

$$a^2 + b^2 = c^2$$

It is important to note that, in keeping with all theorems, the Pythagorean theorem states both a condition and a conclusion. It does not simple state that $a^2 + b^2 = c^2$; the condition of this relationship is that it occurs in a right triangle.

Many math students today who are algebraically-minded do not realize that when we refer to "the square of the hypotenuse," c^2, we are actually referring to a square or more specifically the area of a square. In terms of area, the theorem states that the sum of the area of the two squares, each built based on the sides of a right triangle, is the same as the area of the square based on the hypotenuse. Geometrically, the Pythagorean Theorem can also be understood in terms of *area*.

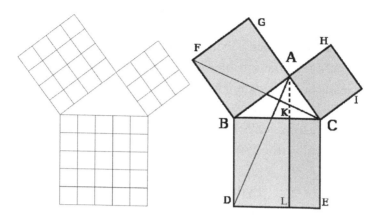

Since there are an infinite number of right triangles, there are an infinite number of values that make the equation $a^2 + b^2 = c^2$ true. However, people have been most intrigued by *whole number solutions* to the equation $a^2 + b^2 = c^2$. For instance, $3^2 + 4^2 = 9 + 16 = 25 = 5^2$, so 3-4-5 provides an integer solution: $3^2 + 4^2 = 5^2$. We call 3-4-5 a *Pythagorean Triple*.

**Plimpton 322 showing Pythagorean Triples
Babylonian; c.1800 BC**

There are many Pythagorean Triples: 5-12-13, 6-8-10, 7-24-25. We find these triples highlighted in historical documents from many cultures dating as far back as ancient Babylon. These triples are not just mathematically interesting; they are actually practically helpful. Why? Because the *converse* of the Pythagorean Theorem is also true.

Pythagorean Theorem:
Right Triangle *implies* $a^2 + b^2 = c^2$
Converse of the Pythagorean Theorem
$a^2 + b^2 = c^2$ *implies* Right Triangle

Practically, this means that if we can find three numbers where $a^2 + b^2 = c^2$, then we can build a right triangle. This property became historically important in marking off land boundaries. Let's say we have a rope that is $3 + 4 + 5 = 12$ meters long, and we mark it every meter. Since 3-4-5 satisfies $a^2 + b^2 = c^2$, we can stretch the rope out in the appropriate lengths, and we will form a right triangle. This process *guarantees* that we construct a physical right triangle when using Pythagorean Triples. This property is so useful it is still used in construction today.

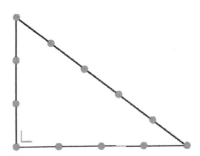

Extension

There are many integer solutions to $a^2 + b^2 = c^2$. For centuries, mathematicians had further related questions: Are there integer solutions to $a^3 + b^3 = c^3$? $a^4 + b^4 = c^4$? $a^5 + b^5 = c^5$?, etc? It turns out that no matter how hard people try, they *cannot* find integer solutions to these equations. An ancient Greek mathematician Diophantus posed another related question. While studying Diophantus' work in 1637, the great French mathematician Pierre de Fermat claimed to have proven that there are no solutions. However, instead of writing the proof itself, he wrote a note in his Diophantus text, "I have discovered a truly marvelous proof of this, which this margin is too narrow to contain."

Did Fermat really develop the proof? No one knows, but most mathematicians doubt it. The problem of confirming Fermat's conjecture stood until 1995 when Princeton mathematician Andrew Wiles finally proved the so-called "Fermat's Last Theorem."

Beautiful Proof?

In this section, you are going to examine five proofs of the Pythagorean Theorem. Why look at more than one proof? Because we need to consider a bigger question. A proof is a convincing mathematical argument that, within given conditions, a claim or conclusion must be true. In the case of the Pythagorean Theorem, the claim is $a^2 + b^2 = c^2$ under the condition of a right triangle with side lengths a and b and hypotenuse length c. Our bigger questions to consider are these: Is one proof more beautiful than another? Is one proof more artistic than another? Is developing a proof an art?

For each proof, your task is to strive to understand what each proof is saying and how it is saying it. Only then will we be able to consider those important questions from an informed perspective. Since the condition in the theorem is that we must have a right triangle, we must first identify a general right triangle. Our first case gives us four right triangles with sides length a-b-c. Notice that these are *arbitrary* triangles, meaning they are not triangles with any *particular* lengths. This property will be true for all of the proofs we consider.

Here is one additional note about *reading* proofs: we cannot read proofs like we read stories. We read stories smoothly and consistently. We usually do not pause to think mid-story because the author generally clears up any doubt or confusion with more explanation. When reading a proof, we must pause, consider, and understand each line before moving to the next line. Every line takes some mental work. You need to ask, "Where did that come from?" and "Why did they do that?" and "How did they do that?" A six-line proof may take the same amount of time to read as six paragraphs of a story. Once, my friend and I spent two hours reading four lines of a proof. We really wanted to *understand* the proof. Understanding develops mathematical insight. With that said, here are the proofs. You may just learn to enjoy them.

Proof 1: Equal Areas

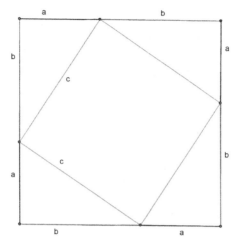

The area of the large square must equal the area of the large square. What we will do is express that area in two different ways: (1) as the area of the large square and (2) as the sum of the areas of the four triangles and the small square that make up the large square.

1. *Area of large square = Area of large square*

2. *large square = 4 triangles + small square*

Use appropriate area formulas

3. $(a + b)(a + b) = 4\left(\frac{1}{2} \times a \times b\right) + (c \times c)$

Multiply

4. $a^2 + ab + ba + b^2 = 2ab + c^2$

Simplify

5. $a^2 + 2ab + b^2 = 2ab + c^2$

Subtract $2ab$ from both sides of the equation.
6. $a^2 + b^2 = c^2$

Proof 2: No Algebra

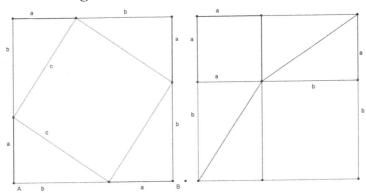

In the image, the two encompassing, outside squares have equal areas (the lengths of their sides are a+b).
1. The area of the square on the left consists of 4 congruent right triangles and a large square.
2. The area of the square on the right consists of the same 4 congruent right triangle, a medium square, and a small square.
3. Since the area of the two starting squares is equal and the 4 triangles in each are equal, then the area of the *leftover parts* inside the original squares must be equal.
4. Starting square minus 4 triangles = Starting square minus 4 triangles
5. So the area of the large square = the area of the medium and small squares put together. 6. Large red square: c^2. Small and medium squares: $a^2 + b^2$.
7. So $a^2 + b^2 = c^2$

Proof 3: Bhaskara's "Behold" (12th Century Mathematician from India)

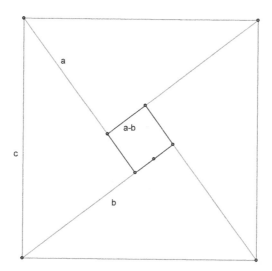

The area of the whole big square can be found two ways.

1. Using 4 triangles of $\frac{1}{2}ab$ and one square of $(a-b)^2$
2. Using the side of the whole big square, the area is c^2.

These two calculations of the area of the big square must be the same:

3. $\left[4\left(\frac{1}{2}ab\right) + (a-b)^2\right] = c^2$

Multiply:
4. $[2ab + a^2 - ab - ba + b^2] = c^2$

Simplify
5. $[2ab + a^2 - 2ab + b^2] = c^2$

6. $a^2 + b^2 = c^2$

Proof 4: Similar Triangles

Reminder: Triangles that are *similar* to each other have *proportional sides*.

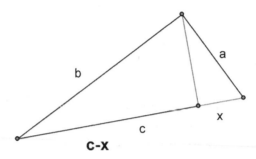

There are 3 similar right triangles in this image: Small, Medium, and Large. Note that the long side of the Large triangle is length "c" and is broken up into two parts: "c-x" and "x."

Using the Small and Large triangles, we can set up the proportion using this hypotenuse-to-short-side ratio: $\frac{a}{x} = \frac{c}{a}$

Using the Medium and Large triangles, we can set up the proportion using the hypotenuse-to-long-side ratio: $\frac{b}{c-x} = \frac{c}{b}$

Cross-multiplying the first proportion results in $a^2 = cx$.
Cross-multiplying the second proportion results in $b^2 = (c-x)(c) = c^2 - cx$.

We then add the two equations:
On the left side, the result is $a^2 + b^2$.
On the right side, the result is $cx + c^2 - cx$.

The sum of the left equals the sum of the right: $a^2 + b^2 = cx + c^2 - cx$.
So $a^2 + b^2 = c^2$.

Proof 5: President Garfield's proof

For President Garfield's (yes, the 20th President of the United States) proof, you will have to recall an extra area formula: The area of a trapezoid is $\frac{1}{2}h(b_1 + b_2)$ where h is the height of the trapezoid and b_1 and b_2 are the lengths of the parallel bases.

Notice that there is a right triangle with side lengths a-b-c involved again. We can find the area of Garfield's trapezoid in two different ways:
1. We can use the area formula for a trapezoid.
2. We can add the areas of the 3 triangles.

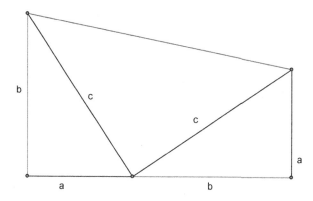

1. Using the trapezoid area formula and the image below, the bases of the trapezoid are lengths a and b while the height of the trapezoid (the bottom edge in this case) is $a + b$. Substituting these values into the formula gives us:

$$\text{Area} = \frac{1}{2}h(b_1 + b_2) = \frac{1}{2}(a+b)(a+b).$$

2. Then we using the three triangles and the triangle area formula:

$$\text{Area} = \frac{1}{2}ab + \frac{1}{2}ab + \frac{1}{2}c^2$$

You can probably anticipate where this is going again! Since the two area expressions are now set and equal, we just need to apply basic algebra:

3. $\frac{1}{2}(a+b)(a+b) = \frac{1}{2}ab + \frac{1}{2}ab + \frac{1}{2}c^2$

Multiply and Simplify

4. $\frac{1}{2}(a^2 + 2ab + b^2) = ab + \frac{1}{2}c^2$

5. $\left(\frac{1}{2}a^2 + ab + \frac{1}{2}b^2\right) = ab + \frac{1}{2}c^2$

6. $\frac{1}{2}a^2 + \frac{1}{2}b^2 = \frac{1}{2}c^2$

7. $a^2 + b^2 = c^2$

Something to Think About

As you read the five proofs, did you find one nicer than another? Most people do. For various reasons, people usually prefer one or two over the others. Which ones did you prefer? Why? Can you put your reasons into words?

Here is a related but more general question. Are some arguments more beautiful than others? Maybe one is simpler, more concise than another. Maybe one is "slicker" than another; it takes a clever and unexpected twist.

Mathematician G.H. Hardy said, "A mathematician, like a painter or poet, is a maker of patterns. If his patterns are more permanent than theirs, it is because they are made with ideas." Bertrand Russell, another mathematician, also said, "Mathematics, rightly viewed, possesses not only truth, but supreme beauty — a beauty cold and austere, like that of sculpture." Do you think of doing mathematics as an art? Is developing a proof more of a science or an art? If you were in charge of a university, and mathematics was not in its own department, would you put math in the "Art Department" or the "Science Department" or another department?

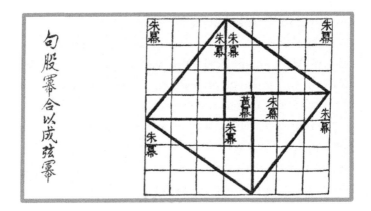

Covering the Reading

1. Determine which of the following are Pythagorean Triples: 9-40-41; 8-15-17; 6-10-12; 10-24-26; 7-32-36.

2. Use the Pythagorean Theorem to find:
    a. The length of the hypotenuse on a right triangle with sides length 10 and 16.
    b. The length of the leg of a right triangle with side length 14 and hypotenuse 21.
    c. The length of the sides of an isosceles (2 sides equal) right triangle with hypotenuse length 12.

3. Explain what the Pythagorean Theorem means in terms of area.

4. Summarize each of the five proofs in a sentence or two. For example, "In proof x, we compare [this-with specifics] to [that-with specifics] and then use algebra to show $a^2 + b^2 = c^2$.

Problems

5. Sketch a right triangle with side lengths 6-8-10 (you pick the units--centimeters, ¼ inches, etc). Draw the "squares" off the sides of the triangle with areas 36, 64, and 100. Use shading or color to show that $6^2 + 8^2 = 10^2$.

6. Does the essence of the Pythagorean Theorem only hold for squares? That is, if you have a right triangle, would the areas of the semi-circles built off each side have the same relationship? In other words, is the sum of the areas of the semi-circles of the two sides equal to the sum of the area of the semi-circle of the hypotenuse? Test the question using a 3-4-5 right triangle.

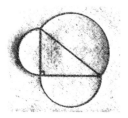

7. Explain how the image below is a picture proof of the Pythagorean Theorem.

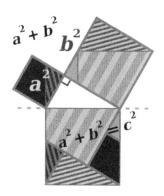

http://commons.wikimedia.org/wiki/File:Academ_Jigsaw_puzzle_for_the_Pythagorean_theorem.svg

8. Which of the five proofs do you prefer? Why? What is it about that proof that you like? What is it about the other proofs that you do not like?

9. As you have now experienced, an "argument" in mathematics has nothing to do with being angry at someone. An argument is a way of proving or convincing. Some arguments from this chapter may be seen as better, sweeter, more convincing than others. Can you think of an argument from outside mathematics that you thought was a beautiful argument or maybe a beautiful refutation of an idea? (A refutation is an answer to a point or argument that disproves that point.) Identify that argument and explain what made it so beautiful to you.

    Finally, consider what criteria make an argument beautiful. What criteria make an argument annoying or ugly or unattractive? Develop a list for each. Are these personal criteria or universal ones?

10. Consider Paul's argument for salvation by faith alone in the book of Galatians. What are some strengths of the argument? How does his argument compare to the lists you made in problem 9?

Part 2
Reason

Chapter 2.1
Logical Reasoning

The skills required by logic puzzles engage the mind in a synthetic or contrived way that transfers easily to real-life settings. We all use logic in our daily lives. "If I do this, then I can't do that." "If I do this, then that will happen. If that happens, then there will be bad consequences." We use logic in many small as well as significant decisions. While logic is sometimes insufficient to help us decide the best course of action, logic, or logical reasoning, is quite helpful and necessary to decision making. Not surprisingly, mathematicians have been studying logic for thousands of years.

Warm-up Activity

Try this logic puzzle. Feel free to use the table to help you. Do not simply solve the puzzle; reflect on and be ready to explain why you come to each conclusion.

In a certain school, the positions of principal, teacher, and aide are held by Franklin, Jefferson, and Washington, though not necessarily respectively.
1. The aide, who was an only child, earns the least.
2. Washington, who married Franklin's sister, earns more than the teacher.

What position does each person fill?

	Principal	Teacher	Aide
Franklin			
Jefferson			
Washington			

Concept Development

To solve this puzzle, you needed to use several different kinds of logic. We will examine those here both casually and formally. You may have arrived at your conclusions in the order presented here; if so, you may want to read this section in a different order. Simply compare the conclusions to your own and start with that explanation.

	Principal	Teacher	Aide
Franklin			
Jefferson			
Washington		X	

Conclusion A: Washington is *not* the teacher. Reason: Washington *earns more* than the teacher.

	Principal	Teacher	Aide
Franklin			
Jefferson			
Washington		X	X

Conclusion B: Washington is *not* the aide. Reason: Washington *earns more* and the aide *earns the least*.

	Principal	Teacher	Aide
Franklin			
Jefferson			
Washington	√	X	X

Conclusion C: Washington is the principal. Reason: Conclusions A and B eliminate the other options.

	Principal	Teacher	Aide
Franklin	X		
Jefferson	X		
Washington	√	X	X

Conclusion D: Franklin and Jefferson are *not* the principal. Reason: Washington is the principal, and so the other two men can't be the principal.

	Principal	Teacher	Aide
Franklin	X		X
Jefferson	X		
Washington	√	X	X

Conclusion E: Franklin is *not* the aide. Reason: Franklin has a sister, and the aide is an only child.

	Principal	Teacher	Aide
Franklin	X	√	X
Jefferson	X		
Washington	√	X	X

Conclusion F: Franklin is the teacher. Reason: Because of conclusions D and E, Franklin is neither the principal nor the aide.

	Principal	Teacher	Aide
Franklin	X	√	X
Jefferson	X	X	
Washington	√	X	X

Conclusion G: Jefferson is *not* the teacher. Reason: Franklin is the teacher.

	Principal	Teacher	Aide
Franklin	X	√	X
Jefferson	X	X	√
Washington	√	X	X

Conclusion H: Jefferson is the aide. Reason: By conclusions D and G, Jefferson is neither the principal nor the teacher.

    We solved the puzzle and explained how we found the answers we did. We used reason to solve the puzzle, and our conclusions fit the starting conditions. This fun little game actually forced us to use several formal rules of logic, but without naming them. After a brief introduction to four rules of logic, we'll consider the situation more formally.

4 Rules of Logic

One of the perks of studying logic is that we get to use symbols to simplify the explanations of these four basic rules. *Conditional statements* can be rewritten symbolically. Consider this conditional statement:

$$\text{If it is cold, then I wear a coat.}$$

To write this statement in symbolic logic, we use a variable for each part of the sentence:

$$\text{Let P: It is cold}$$
$$\text{Let Q: I wear a coat}$$

The conditional statement can now be rewritten as

$$P \rightarrow Q.$$

Two other symbols are helpful. First, $\sim P$ means "not P." In this case, $\sim P$ means it is *not true* that "it is cold." Informally, we would say, "It is *not* cold." Second, the symbol $\therefore$ means "therefore."

These symbols let us consider other "if-then conditional statements.

Let the original conditional statement be $P \rightarrow Q$.
The *converse* of the statement is $Q \rightarrow P$.
The *inverse* of the statement is $\sim P \rightarrow \sim Q$.
The *contrapositive* of the statement is $\sim Q \rightarrow \sim P$.

Or in words:

> Let the original conditional statement be "If it is cold, then I wear a coat."
> The *converse* is "If I wear a coat, then it is cold."
> The *inverse* is "If it is not cold, then I do not wear a coat."
> The *contrapositive* is "If I do not wear a coat, then it is not cold."

Before we move on, we need to note the *relative* truth of the conditional statements. Suppose it *is* true that "If it is cold, then I wear a coat." Is the converse of the statement true? Is it true that "If I wear a coat, then it is cold"? Not necessarily. There could be other reasons for me to wear a coat. Maybe it's raining.

Rule 1: Modus ponens. Modus ponens (MP) follows this logical structure: Suppose it is true that $P \rightarrow Q$ and suppose P is true, what *must* we logically conclude? We must conclude that Q is true. In the case of our conditional statement above, if it is true that "If it is cold, then I wear a coat" and if it is true that "It is cold," then we must conclude that "I wear a coat." More formally:

$$P \rightarrow Q$$
$$P$$
$$\therefore Q$$

Rule 2: Modus tollens. Modus tollens (MT) follows this logical structure: Suppose it is true that $P \rightarrow Q$ and suppose $\sim Q$ is true (informally, Q is false), what *must* we logically conclude? We must conclude $\sim P$ is true (informally, P is false). Now let's apply this to our conditional statement above: if it is true that "If it is cold, then I wear a coat," and if it is true that "I do not wear a coat," then we must conclude that "It is not cold." More formally:

$$P \to Q$$
$$\sim Q$$
$$\therefore \sim P$$

Rule 3: Law of contrapositive. The law of contrapositive (LC) is a lot like MT. However, the conclusion we reach with LC is not a statement but a new conditional statement. In other words, if one conditional statement is true, then another conditional statement is true. Here is the logical structure: Suppose it is true that $P \to Q$. Then it is also true that the contrapositive $\sim Q \to \sim P$ is true. In words, if it is true that "If it is cold, then I wear a coat," then it is also true that "If I do not wear a coat, then it is not cold." More formally,

$$P \to Q$$
$$\therefore \sim Q \to \sim P$$

Rule 4: Law of syllogism. The law of syllogism (LS) is the easiest to understand. Basically, if P implies Q and if Q implies R, then in essence P implies R. More formally,

$$P \to Q$$
$$Q \to R$$
$$\therefore P \to R$$

Each of these rules of logic has some complexities you can study in a full course just in logic! But for our purposes, these simple versions of the rules presented above suffice. Let's now return to the logic puzzle from the beginning of this chapter and consider how the rules are applied. Most of the statements in our logic puzzle use "or" statements and thereby complicate the logic. However, here are two simple examples of how we used MT and MP.

"**Conclusion A: Washington is *not* the teacher. Reason: Washington earns more than *the teacher*.**"

In this scenario, we used MT:

If Washington were the teacher, then Washington would earn what the teacher earns. ($P \rightarrow Q$)

Washington does not earn what the teacher earns. ($\sim Q$)

Therefore, Washington is not the teacher. ($\therefore \sim P$)

"**Conclusion D: Franklin and Jefferson are *not* the principal. Reason: Washington is.**"

In this scenario, we use MP:

If Washington is the principal, then no one else can be the principal. ($P \rightarrow Q$)

Washington is the principal. (P)

Therefore, no one else can be the principal. ($\therefore Q$)

Logical Argument

These basic rules of logic set rules for a game that can be played called *mathematical proof*. In a mathematical proof, we take what is *given* and then use *logic* to establish the validity of a *conclusion*. Whether in algebra, geometry, or any other field of mathematics, this method of proof or logical argument provides the basis for establishing new theorems. In this section, we will construct proofs that are completely symbolic using the rules of logic. As you read through the following example, be prepared to frequently flip back to reference MP, MT, LC, and LS as you follow these proofs. Furthermore, take time to carefully read and understand how each statement is established.

Example 1a:

Given:
$P \rightarrow Q$
$Q \rightarrow R$
P
Prove:
R

Statement	Reason
1. $P \rightarrow Q$	Given
2. P	Given
3. Q	Lines 1 and 2 using MP
4. $Q \rightarrow R$	Given
5. $\therefore R$	Lines 4 and 3 using MP

This is a logical proof because the "Given" statements are filtered through logic to establish the "Prove" statement. While this may seem rather mechanical, it is in proof that mathematics emerges as artistry. Often, many different paths can lead to the answer. Mathematicians use *creativity* to establish one of many valid paths. Here is the same conclusion through a different path.

Example 1b:

Given:
$P \rightarrow Q$
$Q \rightarrow R$
P
Prove:
R

Statement	Reason
1. $P \to Q$	Given
2. $Q \to R$	Given
3. $P \to R$	Lines 1 and 2 using LS
4. P	Given
5. $\therefore R$	Lines 3 and 4 using MP

Example 2: Here is one possible path to $\sim S$

Given:
$P \to \sim Q$
$R \to Q$
$S \to R$
P
Prove:
$\sim S$

Statement	Reason
1. $S \to R$	Given
2. $R \to Q$	Given
3. $S \to Q$	Lines 1 and 2 using LS
4. $\sim Q \to \sim S$	Line 3 and LC
5. $P \to \sim Q$	Given
6. $P \to \sim S$	Lines 5 and 4 and LS
7. P	Given
8. $\therefore \sim S$	Lines 6 and 7 and MP

More on Logic

The formal logic and proof described above is the basis for establishing "truth" in mathematics. Algebraic proof, geometric proof, and other mathematical proofs all use this general structure when directly establishing a theorem. In all direct proofs, a set of statements is accepted as the *givens* for the situation. Mathematical proof uses the rules of logic *and* previously established theorems to prove new theorems. Reasoning and logic are the foundation for formal mathematical proof.

While mathematicians use logical reasoning for proofs, logical reasoning extends beyond the field of pure mathematics. Remember the logic puzzle at the start of this chapter? That problem provided a table to help us reason through the conditions, but logical reasoning can extend beyond these purely fabricated, table-based problems. The following logic problem uses logic in the context of personal experience. You know a number of things about light and lightbulbs. Take five or ten minutes to think about the problem before reading on.

Light switch problem: For some unknown reason, there are 3 light switches in the basement, one of which controls an upstairs incandescent light (old, globe-ish lightbulbs with a filament inside. You probably have a few somewhere in your house). The other 2 are wired but don't do anything at all. We know the light bulb works but is off. All 3 light switches are in the "OFF" position. You cannot see the light or the effects of the light from the basement. If you start in the basement, what is the fewest number of times you must go up and down stairs (each direction counts as one) to be guaranteed that you know which switch controls the light? (note: You have no special electrical tools. You will not dismantle the light switches, etc. This is *not* one of those pure lateral thinking puzzles. Use logic and your experience with light bulbs.)

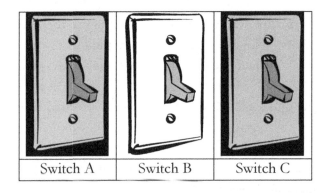

| Switch A | Switch B | Switch C |

Light switch hint and answer: Of course, if you guess correctly, the minimum number of trips *could be* only one. However, the question asks us for the fewest possible number of trips even in the situation where we have the most errors. How many trips are absolutely necessary in every possible scenario for you to be sure which switch operates the light? In our lightbulb situation, the total number of trips that *must* be taken is still one. Why?

Light switch partial solution: For the sake of discussion, let's call the light switches A, B, and C. Here is the partial solution: you fill in the rest. I turn on switch A and wait 5 minutes. Then I turn off switch A. I turn on switch B and go upstairs. I look at and might touch a light bulb. How can I be sure that I know which light switch controls the light?

Something to Think About

This logical reasoning allows us to consider two big ideas. First, we must critique logical reasoning. As we know from earlier in this chapter, a logical argument uses a set of *givens* and employs reason to navigate toward a conclusion. The truth of the conclusion is established based on two things: the reliability of the laws of logic and the reliability of the given statements. I have seen that when people disagree on something significant, both parties often think the other a fool. After all, how can two people both be reasonable and hold desperately different viewpoints? Frequently, I have also found that both parties use valid logic.

In Favor of X	Against X	← Apparent Issue
↑Logic	↑Logic	Both people use reason correctly
↑Logic	↑Logic	
Starting Premise A	Starting Premise B	←Real Difference

Both are literally "reason-able" people. However, they begin with different *givens*, different starting assumptions or different presuppositions. With two different starting points, two people can validly use logical reasoning and conclude diametrically opposed conclusions. Realizing that the issue was the starting points and not the people's reasoning has helped and even caused me to be more compassionate and understanding toward people I strongly disagree with.

Secondly, historically long conversation regarding the relationship between faith and reason. St. Thomas Aquinas' writings are foundational in this discussion within the Church. Some see faith and reason as at odds with one another. Others understand the two to be complimentary. Still others see reason as the only reliable way to come to understand or prove the nature of ultimate reality: God. Examining the historical views on the relationship between faith and reason is worthy of study.

Covering the Reading

1. Use the following conditional statement as a starting point. Write its *converse, inverse,* and *contrapositive.* "If I study hard, then I do well on an exam."

2. Convert the following statements to symbols and identify which rule of logic is being used in this scenario: "If I study hard, then I do well on an exam. I did not do well on an exam. Therefore, I did not study hard."

3. Convert the following statements to symbols and identify which rule of logic is being used in this scenario: "If I study hard, then I do well on an exam. If I do well on an exam, then I pass the class. Therefore, if I study hard, then I pass the class."

4. Give an alternate proof of Example 2.

5. Explain the solution to the light switch problem.

Problems

6. Use a table and reasoning to solve the following logic puzzle. Aquinas, Descartes, and Luther make their living as carpenter, electrician, and plumber, though not necessarily respectively.
    1. The electrician recently tried to get the carpenter to do some work for him, but was told that the carpenter was out doing some remodeling for the plumber.
    2. The plumber makes more money than the electrician.
    3. Descartes makes more money than Aquinas.
    4. Luther has never heard of Descartes.

What is each person's occupation?

7. Fill in the reason for each statement.

Given: $P \rightarrow \sim Q$ $\sim R \rightarrow Q$ $R \rightarrow S$ $\sim S$ Prove: $\sim P$	
Statement	Reason
1. $R \rightarrow S$	
2. $\sim S \rightarrow \sim R$	
3. $\sim R \rightarrow Q$	
4. $\sim S \rightarrow Q$	
5. $\sim S$	
6. Q	
7. $P \rightarrow \sim Q$	
8. $\therefore \sim P$	

8. Prove the case in number 7 in a different way.

9. Classic "Crossing the River" Problem: This problem is found in many different books, written in slightly different ways.
Once there was a man who had to take a dog, a cat, and a rat across a river. But his boat was so small it could only hold himself and one other thing. The man didn't know what to do. How could he take the dog, the cat, and the rat over one at a time, so that the dog wouldn't eat the cat and the cat wouldn't eat the rat? In other words, the dog and cat could never be left alone without the man and the cat and rat could never be left alone without the man.

10. In this chapter, we identified a weakness of logical reasoning: if two people begin with different premises or different assumptions, they can logically come to different valid conclusions. Provide an example of a disagreement on an important issue that fits these criteria.

11. Research and explain one major view on the relationship between faith and reason. Summarize the view and then critique it. What do you agree and disagree with? Why?

Chapter 2.2
Inductive and Deductive Reasoning

Deductive reasoning is pure logic. When we begin with certain premises and logically conclude something, we are using deductive reasoning. We use deductive thinking when we reason, "If A is true then B will logically follow." The art of logical inference allows us to deduce what *must* be the case. In deductive reasoning, we begin with a premise and deduce the logical consequences of the premise.

In contrast to deductive reasoning, *inductive* reasoning is theory-building. Evidence helps us conclude what is *probably* the case. However, more evidence may come to light that makes us modify our previous theory. We collect data, develop a theory, collect more data, and revise the theory.

Consider the following example. When my daughter was a toddler, she saw someone carrying a Chihuahua in her arms. My daughter pointed at the tiny dog and said, "Cat." We corrected her and said, "No. That is a dog." She insisted, "No! Cat!" Prior to that point in her early life, every *small* four-legged furry creature with pointy ears she experienced was a cat while large ones were dogs. She was using inductive reasoning to build the construct of "Cat." However, new evidence eventually forced her to change that construct. She used inductive reasoning to come to her conclusions. In contrast, I might use deductive reasoning this way: I am allergic to all cats. I am not allergic to a Chihuahua. Therefore, a Chihuahua is *not* a cat.

In the previous chapter, we focused on deductive reasoning: pure logic. This chapter will center on inductive reasoning. Try the following activity.

Warm-Up Activity
In each list below, three of the numbers are similar while the fourth is different. Which number in each set is different?

Set 1:	2	$-3/8$	-6	-1
Set 2:	$2/4$	$8/16$	$5/10$	$4/12$
Set 3:	2	$\sqrt{2}$	e	2π

Concept Development
Were you able to argue which number was different in each set? Consider set 1. Which number did you determine was different? Could you argue that 2 is different? Could you argue that $-3/8$ is the number that is different? Each of the sets is designed so that there are at least two possible answers; you could make a hypothesis as to what 3 of the numbers have in common but also argue that three others have something in common. You can develop a theory, but you need more data to support your theory. For example, suppose we added the following two numbers to Set 1: -4 *and* $-5/8$. Of the six numbers, which one is different? You can have a stronger theory about "2" at this point, but you still cannot be certain.

Activity 2

Let's continue to apply inductive reasoning to patterns. For each of the following sequences, write the next three terms. Then rate your level of confidence in your predictions: Very Confident (VC), Somewhat Confident (SC), Neutral (N), and Unconfident (U).

	Sequence	Prediction 1	Prediction 2	Prediction 3	Confidence?
A.	Summer, Fall, Winter, Spring, Summer, Fall, …				
B.	2, 8, 14, 20, …				
C.	$\frac{1}{2}, \frac{1}{4}, \frac{1}{8}, \frac{1}{16}, \ldots$				
D.	A, B, D, G, K, …				
E.	2, 3, 5, 7, 11, 13, …				
F.	Triangle, Square, Pentagon, Hexagon, …				
G.	1, 3, 6, 10, 15, …				
H.	1, 1, 2, 3, 5, 8, 13, …				
I.	J, F, M, A, M, J, J, …				
J.	O, T, T, F, F, S, S, …				

K.	1, 1, 1, 2, 4, 8, 3, 9, 27, 4, …				
L.	1, 10, 11, 100, 101, 110, 111, 1000, 1001, 1010, …				
M.	1, 11, 21, 1211, 111221, …				

In this activity, you used inductive reasoning: you developed a theory related to the sequence and made a prediction. Consider the next list below, which states the next term for each sequence as I intended it. Does each answer below change your prediction 2 or 3? Does it change your confidence? A. Winter; B. 26; C. 1/32; D. P; E. 17; F. Heptagon; G. 21; H: 21; I. A; J. E; K. 16; L. 1011; M. 312211.

Activity 3

We also use inductive reasoning to develop mathematical concepts. Consider the two examples involving algebra and geometry. You probably already know the answer to each of the questions. However, you should be able to explain how we can use inductive reasoning in each situation.

3.1 Consider the graphs and equations below:

$y = 2x$	$y = 4x$	$y = -\frac{1}{2}x$	$y = -3x$	$y = \frac{1}{3}x$
What is the impact of the coefficient on x? In other words, in the equation $y = mx$, what is the relationship between m and the graph?				

The responses to this problem from a class of beginning algebra students are often quite varied; they usually involve the steepness of the line or the "slanty-ness" of the graph. But one more step makes is easy for students to realize the impact of *m* and *b* on the graph of the equation $y = mx + b$. This is one example of using inductive reasoning to do mathematics. Here is another example.

3.2 Consider the examples and non-examples below:

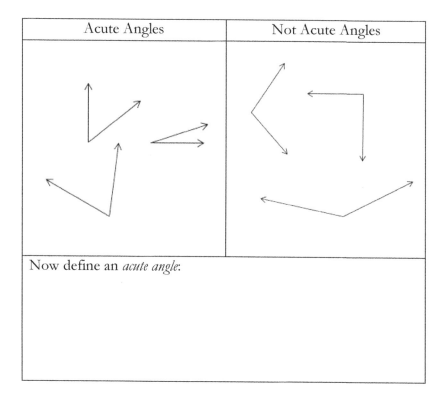

By studying examples and non-examples, we use inductive reasoning to develop definitions and concepts. This is the basic way we learn many things in life as well as one of the foundations for doing and learning mathematics.

Deduction and Induction Working Together

Mathematics that is written down in books or presented at conferences is typically in its logical, deductive final form. If we were to examine a typical advanced mathematics text, we would find definitions, theorems, and proofs. Definitions usually come with a few examples and non-examples to help clarify the explanation. Then, the book presents theorems—mathematical truths—with a logical, deductive proof. However, most people never consider the process through which mathematicians arrived at those conclusions. That process it usually messy and frustrating, but it can also be energizing and fun.

While the product of mathematics is deductive, the process of mathematics is frequently inductive. To put it simply, mathematicians see patterns. This ability see patterns is frequently the mathematician's greatest strength. Many are good with numbers and have strong visual-spatial reasoning skills. However, the ability to recognize a pattern, to generalize the pattern, and to prove that the pattern is consistent is the true power of the mathematician. This sequence of inductive pattern-recognition and generalization, followed by deductive reasoning to prove something is *always* the case, is the actual *process* of doing mathematics.

The professional mathematician seeks to develop new mathematics by recognizing, understanding, and proving something new, but even the casual or serious student of mathematics can participate in and develop the skills of the professional mathematician. We simply need to be watching for patterns, generalizing patterns, and seeking to develop a deductive argument that our theory—our generalization—is consistent.

Activity 4

Try your skills at being a mathematician. Consider the sums of the first n odd numbers. $1, 1 + 3, 1 + 3 + 5, 1 + 3 + 5 + 7, \ldots$. Is there a pattern that emerges in the totals? Examine more cases. Try to build a pattern of the sums of the first n odd numbers. Try to recognize the nature of the sequence.

Sum of the first n odd numbers?
$1 =?$
$1 + 3 =?$
$1 + 3 + 5 =?$
$1 + 3 + 5 + 7 =?$
$1 + 3 + 5 + 7 + 9 =?$
$1 + 3 + 5 + 7 + 9 + \cdots + n =?$

The sums are not just related to odd numbers but also to a different sequence. Look at the sequence independent of the sums of odd numbers. What is the pattern? Can you establish a relationship between the sum of the first n odd numbers and the pattern that you found?

Next, can you prove that the relationship holds? How can you convince someone that the pattern that you observed is *always* the case? This is the hardest part of doing mathematics—proving the certainty of something for an infinite number of cases.

You may or may not be able to do this task. That's okay. How far can you get? Can you keep at the problem over time until you conquer it? That perseverance is the work of doing mathematics. To the mathematician, that work is joy.

Something to Think About

This chapter can help us to consider a few more important questions. First, inductive and deductive reasoning each has its own strengths and weaknesses. Can you identify the strengths and weaknesses of each? Should you trust one more than the other?

The second question is more affective. Do you enjoy the work of doing mathematics? Do you enjoy finding patterns, coming to generalizations, then trying to develop an argument that this is always the case? Do you enjoy one portion of that process more than another?

Covering the Reading

1. Refer to the Warm-Up exercise. Provide a 5th and a 6th number that will clarify which number should be left out of the set.

Set 1:	2	$-3/8$	-6	-1		
Set 2:	$2/4$	$8/16$	$5/10$	$4/12$		
Set 3:	2	$\sqrt{2}$	e	2π		

2. Complete Activity 2:

	Sequence	Prediction 1	Prediction 2	Prediction 3	Confidence?
A.	Summer, Fall, Winter, Spring, Summer, Fall, ...				
B.	2, 8, 14, 20, ...				

C.	$\frac{1}{2}, \frac{1}{4}, \frac{1}{8}, \frac{1}{16}, \ldots$					
D.	A, B, D, G, K, …					
E.	2, 3, 5, 7, 11, 13, …					
F.	Triangle, Square, Pentagon, Hexagon, …					
G.	1, 3, 6, 10, 15, …					
H.	1, 1, 2, 3, 5, 8, 13, …					
I.	J, F, M, A, M, J, J, …					
J.	O, T, T, F, F, S, S, …					
K.	1, 1, 1, 2, 4, 8, 3, 9, 27, 4, …					
L.	1, 10, 11, 100, 101, 110, 111, 1000, 1001, 1010, …					
M.	1, 11, 21, 1211, 111221, …					

3. Complete Activity 3.1 and 3.2.

4. What are the differences between inductive and deductive reasoning?

Problems

5. Explain three instances in which you used (a) inductive and (b) deductive reasoning this week.

6. Mathematics is generally associated with deductive logic while science is generally associated with inductive theory-building. However, both disciplines use both kinds of reasoning. How does each discipline use these kinds of reasoning? Provide some examples.

	Inductive Reasoning	Deductive Reasoning
Mathematics		
Science		

7. Inductive and deductive reasoning each has its own strengths and weaknesses. Can you identify the strengths and weaknesses of each? Should you trust one more than the other?

8. Do you enjoy the work of doing mathematics? Do you enjoy finding patterns, coming to generalizations, then trying to develop an argument that it is always the case? Is there one portion of that process that you enjoy more than another?

9. Make progress on Activity 4.

Chapter 2.3
Cause and Correlation

Cause and correlation are often used interchangeably by the media and today's popular culture; this is a widespread misinterpretation of both terms. A *correlation* is simply a relationship between two variables. However, a correlation does not imply *cause*; in other words, just because two things are related does not mean that one causes the other. For example, we can establish a clear relationship between adult height and adult shoe size; in general, as height increases, shoe size increases. Taller people generally wear larger shoes than shorter people. That is a correlation. However, we could misinterpret this as a *causal* relationship. Suppose I want to be taller. What should I do? Obviously, I should wear bigger shoes. Since big shoes go with tall people, all I have to do is purchase some size 13 shoes, and then I will be over 6 feet tall! Do you see how treating correlation as cause is a problem?

Warm-up Activity

When two variables are correlated, they are correlated either *positively* or *negatively*. In this case, "positive" and "negative" do not mean "good" or "bad." Instead, they specify the nature of the relationship. We say two variables are *positively correlated* if they relate to each other in the same direction: as one variable increases, the other variable increases, or as one variable decreases, the other variable also decreases. For example, almost everyone believes that studying and grades are positively correlated: as studying increases, grades increase. Studying little relates to low grades.

Positive Correlation			
Variable A	↑	↑	Variable B
Variable A	↓	↓	Variable B

On the other hand, variables are *negatively* correlated if they relate to each other in opposite directions: as one variable increases, the other variable decreases, or as one variable decreases, the other variable increases. For example, many people believe that class size and student achievement are negatively correlated: as class size increases, student achievement decreases. A small class size relates to high-achieving students.

Negative Correlation			
Variable A	↑	↓	Variable B
Variable A	↓	↑	Variable B

Now that you better understand correlation, list 4 pairs of variables that you suspect are positively correlated and 4 pairs of variables that you suspect are negatively correlated.

	Positive Correlations				Negative Correlations			
	Variable A			Variable B	Variable A			Variable B
1.		↑	↑			↑	↓	
2.		↑	↑			↑	↓	
3.		↑	↑			↑	↓	
4.		↑	↑			↑	↓	

Concept Development

In addition to the direction of the correlation, we can also consider the strength of the correlation. Looking at a scatterplot often helps us think about the strength of a correlation. The following 5 scatterplots show 5 different kinds of correlations: strong positive, strong negative, weak positive, weak negative, and practically no significance. Can you match the plot with the relationship?

Match the Correlation Type to the Scatterplot

Correlation Type	Scatterplot
1. Strong Positive Correlation	
2. Strong Negative Correlation	
3. Weak Positive Correlation	
4. Weak Negative Correlation	
5. Practically No Correlation	

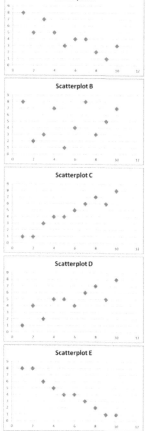

How did you do? Scatterplots C and E demonstrate strong linear correlations. Scatterplot C shows a positive correlation because as the x-axis number increases, the y-axis number increases. Scatterplot E shows a negative relationship because as the x-axis number increases, the y-axis number decreases. Both of the relationships are strong because we can almost draw a straight line directly through all of the points. Scatterplots C and E make it easy to consider the concept of a line-of-best-fit, also called a prediction line. The points in these scatterplots overall lie very tightly to a line. What would be the indication that points overall lie close to a line? There are two possibilities: either there a very few points that deviate from the line or there are many points that deviate a tiny bit. Either way, the total amount of deviation from the prediction line is very small.

Scatterplots A and D also demonstrate correlations. However, these relationships are not as strong. Scatterplot A shows a negative correlation while scatterplot D shows a positive correlation. We can see that these correlations are weaker because although we see a general linear pattern, many values land further away from the line-of-best-fit.

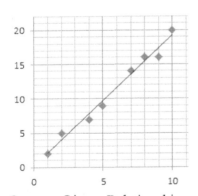

Stronger Linear Relationship
Little deviation from a line-of-best-fit

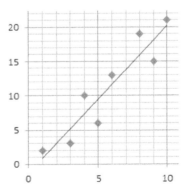

Weaker Linear Relationship
More deviation from a line-of-best-fit

Finally, scatterplot B demonstrates basically no relationship between the variables. Mathematically speaking, there is still a line-of-best-fit. However, the "best fit" is still a very, very bad fit. Imagine that I went into a children's clothing store; I would find something that would fit me "best," but it certainly would not fit me well. That's the type of "best fit" we are talking about in scatterplot B.

Cause and correlation

Since you now have some experience with what a correlation *is*, we are now going to focus on what a correlation *is not*. Let's begin with a few fun, true examples:

1. For a certain segment of the population, there is a positive correlation between height and reading ability.
2. There is a positive correlation between ice cream sales and crime.
3. In a mall setting, there is a positive correlation between hiring temporary workers and sales.
4. There is a positive correlation between number of crimes committed and police presence.

Consider each of these true correlations. Remember that a correlation is just a relationship between two variables. If two variables are correlated, one *may* cause the other, but it is invalid to conclude that one *causes* the other.

If we misinterpret the correlation in example 1 as cause, we would look to taller people as great readers. Conversely, we might take specialized reading courses to help us grow taller. But the correlation in example is true: the "certain population" is children. You have surely noticed that children in kindergarten are both short and do not read well. However, children in third grade are generally taller and read better. Finally, sixth graders are generally much taller and read much better. There is our correlation. Even if we know that the population in question is just children, we could still foolishly misinterpret correlation as cause; we might be more concerned about childhood nutrition to help kids grow bigger and faster in order to help them read better. Conversely, we might hire a private reading tutor for a child who is currently shorter than his or her peers. This example is obviously silly because the actual cause of the correlation is obvious. However, we frequently fall for the cause/correlation trap without noticing it.

Ice Cream Sales   **Crime**

In the second example, we observe a positive correlation between ice cream sales and crime: as ice cream sales increase, crime increases. Personally, I love ice cream, but should I feel badly about my contribution to crime? The correlation between ice cream and crime ties back to a different variable. Crime and ice cream sales both increase in the summer and decrease in the winter. Hence, the correlation between crime and ice cream sales is real, or true, even though the two variables have nothing to do with one another. Again, this is a fun example. However, these types of correlations have resulted in legislation. It is not difficult to imagine a certain congressman or woman latching onto a correlation like this and proposing a bill to ban the sale of ice cream to help fight crime. Of course, this bill would be accompanied by the demonization of soft-serve in the popular media in an effort to raise public awareness. In preparation for the anticipated ban on ice cream, backlash bloggers would propose an underground network of ice cream distribution. After the bill passes, certain states would propose a referendum to legalize ice cream in their state in the November election. Ultimately, the Supreme Court would be required to rule if banning the sale of ice cream violates some constitutional rights. Hopefully this scenario would not go this far, but the misinterpretation of correlation and cause regularly yields this kind of effects.

Now that you have negotiated the first two examples, the third example—increased retail sales and hiring temporary workers—should not be too difficult to see through. When do retail stores hire temporary workers? When do retail stores do the most business? We can thank the holidays for this correlation.

Can you see the error in misinterpreting the fourth example as causation? If police presence and number of crimes are positively correlated, would it not make sense to fire all of the police officers in every city? That would eliminate crime, right?

Try it: The misinterpretation of correlation as cause can lead to some disturbing conclusions. Reflect on each of these and identify the misinterpretation.
  A. There is a strong correlation between people carrying umbrellas and rain. I should not carry an umbrella so that is does not rain.
  B. There is a strong correlation between teams winning championships and parties in the city streets. My city should pour out into the streets and have a celebration when our team is in the finals.

C. There is a positive correlation between the sales of "organic foods" and autism. I should stop feeding my children organic foods.
D. Philadelphia is one of the fattest cities in the United States, and Minneapolis is one of the healthiest. I should more to from Philadelphia to Minneapolis to lose weight.

Most of the examples above are obvious, maybe even silly, violations of the cause/correlation error. However, these errors are rampant in the media. Let's look at two articles from CBS News, October 11, 2012 (a very normal day): "Impulse Shopping May Contribute to Obesity" and "Eating Chocolate May Help You Win Nobel Prize."

Impulse Shopping May Contribute to Obesity. In this first article, the authors take a study that establishes a correlation between impulse shopping and obesity and present the relationship as possibly causal. They imply that one causes the other. Can you think of a third variable that would be characteristic, possibly a cause, of someone who is an impulse shopper and is also obese?

Eating Chocolate May Help You Win Nobel Prize. The second article is a bit more fun. As a confessing chocoholic, I am enticed by the chocolate-Nobel Prize correlation. This article establishes the correlation between traditionally well-known chocolate-producing nations and Nobel Prize winner-producing nations. Has the article uncovered a causal relationship, or is this another correlation misinterpreted as cause? How do you explain the relationship? Personally, I think I would rather interpret it as cause; the next time my doctor challenges me to cut back on the chocolate, I can tell her that I cannot because I am working toward a Nobel Prize.

The problem we face in popular culture related to the correlation-misinterpreted-as-cause should not surprise us given the value of media hype. The path usually follows this type of progression:

> A researcher finds a weak but statistically significant correlation between two variables, A and B, with certain specific conditions C, D, and E for a specific populations F and G.

> The researcher's institution publishes an article highlighting the correlation between A and B.

> The large news organizations pick up the article and claim that researchers say that A causes B.

> Internet chatter picks up on the article and develops a few conspiracy theories. Readers are convinced that they "knew" that they were trying to kill us.

> Cable news takes the causation as fact and spends hours debating what A causing B will mean for the current presidential administration.

> The evening news advertises and runs a special on the dangers of A.

↓

> The original researcher comes home to find his family throwing away all of the A in the house in order to protect them from B.

Something to Think About

I often wonder why we are so susceptible to the correlation-misinterpreted-as-cause error. Are we gullible or just mis-educated? Do we as humans have a tendency to establish causes or more precisely causal linkages between variables? If so, is the tendency so strong that it can overpower simple reason?

Perhaps the issue has more to do with the way we and our media feast on sensationalism; the more bizarre or the more sensational the relationship, the more we want to revel in it. What characteristics of being human lead us toward this error? Should we do something about those characteristics?

Covering the Reading

1. Describe two variables that you would expect have:
    a. A strong, positive correlation.
    b. A strong, negative correlation.
    c. A weak positive or negative correlation.
    d. Practically no correlation.

2. Use words to describe the relationship pictured here:

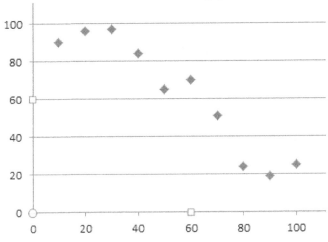

3. Respond to the "Try It" in the chapter: The misinterpretation of correlation as cause can lead to some disturbing conclusions. Reflect on each of these and identify the misinterpretation.

A. There is a strong correlation between people carrying umbrellas and rain. I should not carry an umbrella so that is does not rain.
B. There is a strong correlation between teams winning championships and parties in the city streets. My city should pour out into the streets and have a celebration when our team is in the finals.
C. There is a positive correlation between the sales of "organic foods" and autism. I should stop feeding my children organic foods.
D. Philadelphia is one of the fattest cities in the United States and Minneapolis is one of the healthiest. I should move to from Philadelphia to Minneapolis to lose weight.

Problems

4. For each of the following pairs of variables, determine if you believe there is:
   1. A strong, positive correlation.
   2. A strong, negative correlation.
   3. A weak, positive correlation.
   4. A weak, negative correlation.
   5. Practically no correlation.

Explain your conclusions.
   A. Hat size and shoe size
   B. Neck pain and text messages written
   C. Candy consumption and tooth decay
   D. Caffeine consumption and overall health
   E. Studying and grades
   F. Exercise and vegetable consumption
   G. Number of types of technology and number of pirates (yes, the "Arrr, matey" type)

5. Explain what is wrong with the conclusions of this article:
Headline: Firefighters are the Biggest Polluters!
There is growing evidence that among all people in any given town or city, firefighters are the biggest polluters. A recent study conducted by the National Association of Wasteful Spending found a very strong correlation between the number of firefighters in a given public area and the level of carbon emissions in the vicinity. Both small-town and city mayors have asked their respective city councils to cut funding of fire companies in an effort to go green. Many environmental groups support the measures. However, some do not think that the step of cutting funding goes far enough. A Washington, D.C., area think tank has asked congress to consider a constitutional amendment that prohibits the training and harboring of any former or suspected firefighters. It is likely that the United Nations will eventually need to get involved in the controversy.

6. Find an article that misinterprets correlation as cause.
   A. Identify the correlation.
   B. Identify the cause implied by the article.
   C. Identify a variable that may be the causal explanation.

7. In this chapter, we considered the human susceptibility to the correlation-misinterpreted-as-cause error. We offered several explanations: humans being gullible, people being mis-educated, a human tendency toward establishing causal linkages between variables, and the sensationalism of our culture and media. What characteristics do we as humans possess that predispose us to this error? Which of those characteristics are part of our being made in "the image of God"? Which are products of the fall? Provide a reflective explanation specifically addressing the "correlation misinterpreted as cause" error.

Chapter 2.4
Deception in Statistics

Darrell Huff's 1954 book *How to Lie with Statistics* is not a book designed to do help the reader do what the title implies. Instead it is the classic in thinking about how we are deceived by statistical presentations and practices. Whether intentionally or unintentionally, people regularly use statistics in such a way that lead us to believe something that is not true, is not completely true, or is not as extreme as it appears. In a culture like ours, one that is constantly looking for the next sensational news story, people tend to lack both the desire and the skills to think critically about the statistics they are fed.

Warm-up Activity

Think back to what you have learned about statistics. For the data set below, find the arithmetic mean (the average) and the median. Then, create one graphical representation of the data: a bar graph, histogram, pie chart, stem-and-leaf plot, etc.

```
    17      21      33      42      19
       08      28      24      32      32
    31      41      14      37      19
       15      23      29      35      31
```

Concept Development

In this chapter, we will examine several glaringly deceptive statistical practices, although there are many more out there. But we are also manipulated using statistics in more subtle ways that we do not notice as easily. The goal of this chapter is to help you become more savvy and cunning in your dealing with statistics and statistical presentations. Let's begin by looking at some of the most common misuses of statistics.

Graphical Deception

The Scale Trick: These three graphs present the exact same data but might tell three different stories. Look at the three graphs and decide what story each one encourages. What headline would you attach to each graph?

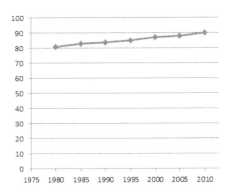

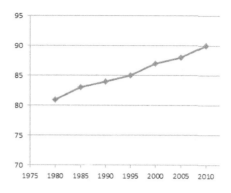

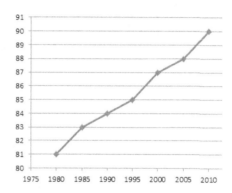

The first graph would usually be accompanied by a statement that there is "virtually no growth" over the 30 year period in question. The second graph could easily be accompanied by the statement that there is "slow but steady growth" while the third graph would probably come with words like "extreme," "severe," or "rampant." And yet all three graphs present the same exact data. So what is different? The way the people making the graphs use the *scale* shows a different story each time.

The Area/Volume Trick: Consider the two charts below. Again, they both show the same data but in different ways. These charts reflect no particular data about Philadelphia and New York.

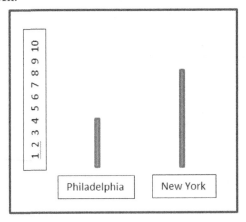

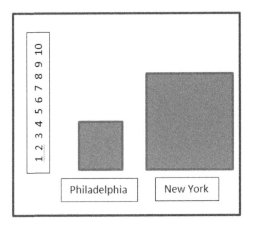

In both cases, New York ranks two times as high as Philadelphia. However, in the second chart, the area of the square representing New York is four times as large as the one for Philadelphia. People who use this trick take advantage of the relationship between length and area to make the difference in data seem more severe than it actually is.

Let's examine the mathematics behind this trick more closely. Consider two segments: 1 inch long and 2 inches long. We would say the lengths are in a 1:2 ratio with one another.

However, if we were to make two squares one square is 1 inch on each side while the other is 2 inches on each side. The area of the small square is 1 square inch, and the area of the larger square is 4 square inches. The areas are in a 1:4 ratio with one another.

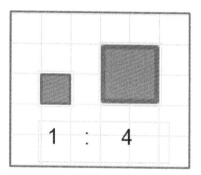

This difference in length ratios compared to *area* ratios is often used to deceive us into thinking a difference is much bigger than it really is. This difference is frequently magnified even further by using the length ratios compared to the *volume* ratios. A cube with side lengths 1 inch long and a cube with side lengths 2 inches long each respectively yield volumes of 1 cubic inch and 8 cubic inches. The volumes are in a 1:8 ratio with one another.

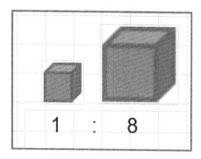

Graphs that contrast area or volume instead of length are often used to convince us something is more significant that it really is.

Relative Position: In this deception, small differences are made to look larger using perspective. In a perspective drawing, things that are "further away" appear smaller. The key is to twist the graph in such a way that whatever you want to appear largest is placed "closest" to the viewer. In the data below, the values represented are 4.8, 5.0, and 5.2 for California, Florida, and New York respectively. In the second graph, New York is twisted closer to appear even larger than the data actually bear out.

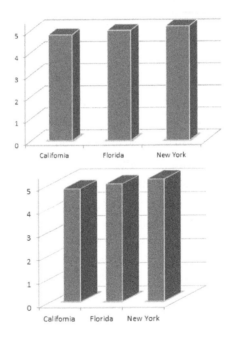

Here are some other ways to graphically "twist" data. The images below employ all three of the tricks described above to some extent. Almost no difference can easily be made to look like fairly severe differences.

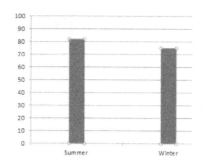

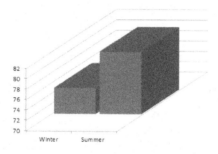

Population Density Trick: Let's look at one final graphical trick: the use of land areas to represent people or some other people-related activity. Here is how the trick works:

At the time of the writing of this book, the population of the United States was about 310,000,000. According to the U.S. Department of Health and Human Services, about 12,800,000 or 4.1% of the population was on welfare. This number can be represented in two different ways: as the population of New Jersey and Connecticut or as the population of Montana, North Dakota, South Dakota, Wyoming, Nebraska, Kansas, Nevada, and New Mexico. Which map shows welfare as more pervasive?

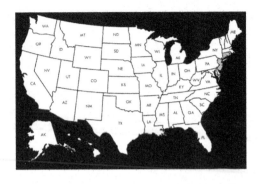

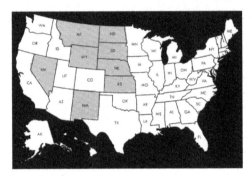

You have seen these types of maps before. Where? Mobile service providers! A classic example of this phenomena discussed in the media related to Verizon and AT&T's old 3G coverage. This map was examined in a legal battle. (if you are curious about this case, continue your reading here: http://www.digitaltrends.com/mobile/verizon-tells-att-that-the-truth-hurts/ and http://www.cultofmac.com/22525/verizon-and-att-stop-squabbling-drop-their-theres-a-map-for-that-lawsuits/)

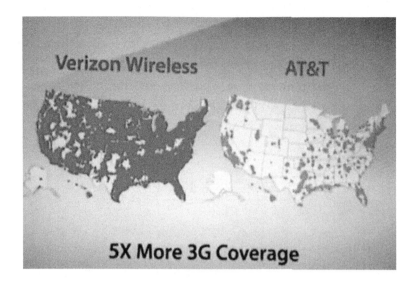

This map clearly shows that Verizon has five time—5X—more 3G coverage by *land area*. However, we have to consider that since AT&T's coverage *completely* covers *all* of the major metro areas, AT&T's coverage probably covers far more than 1/5 of the actual calls people attempt to make. The northeast megalopolis from Washington, D.C., to Boston alone hosts about 17% of the U.S. population in 2% of the land area. This trend continues for many of the population centers covered by AT&T. When we put all this data together, we realize that although Verizon may cover 5X the area, they may cover as little as 1.5X the calls. Is this practice false advertising? The related lawsuits were eventually dropped. What do you think?

Numerical Deception

Now that you better understand graphical deception, let's take a look at some numerical deception tactics. Media reports often focus on two statistics in a data set: the mean and the median. We are easily led astray by these types of reports because of our lack of understanding of the differences between these two statistics. As you may or may not recall from your studies in statistics, the *mean* and *median* are two different "measures of central tendency." In other words, they are two different ways to describe the nature of a group in general.

Here is a quick review: to calculate the mean or average of a set of numbers, you take the sum of the terms and divide by the number of terms. Given the data set 13, 22, 28, 17, and 25, the average is:

$$\frac{13 + 22 + 28 + 17 + 25}{5} = 21$$

The median is simply the middle number of the set when put in order from least to greatest or greatest to least. If we reorganize the numbers 13, 22, 28, 17, and 25 in ascending order and select the middle number, we get a different result: 22.

13 17 **22** 25 28

How can we be deceived by numbers that are so simply defined? There are two main reasons: our understanding of (1) how the mean and median are impacted by numbers and (2) how using the mean and median on different occasions can sway our thinking.

Let's work with the data set from above to better illustrate how the mean and median can be impacted by numbers.

13 17 **22** 25 28

What happens if we change one number? What is the impact of changing the 28 to 68? The *median* will still be 22:

13 17 **22** 25 68

However, the *mean* has changed dramatically!

$$\frac{13 + 17 + 22 + 25 + 68}{5} = 29$$

Notice that the mean is significantly impacted by extreme values while the median is not.

Let's use another example to help us think through this difference between mean and median. Consider the home values in a certain area. We will express these values in hundred-thousands: that means 250 represents a $250,000 home while 2300 represents a $2,300,000 home.

Home Values (in 1,000's)				
215	355	275	405	310
3500	260	285	325	445
190	700	280	260	400

190, 215, 260, 260, 275, 280, 285, **310**, 325, 355, 400, 405, 445, 700, 3500

The median for this set of home values is 310.
The mean for these data is 547.

$$\frac{215 + 355 + 275 + 405 + 310 + 3500 + 260 + 285 + 325 + 445 + 190 + 700 + 280 + 260 + 400}{15}$$

$$= 547$$

Let's recap: the median value of a home in this area is $310,000 while the mean value of a home is $547,000. This is quite a significant difference! Why are the mean and median so different? The answer is that the one *very* expensive home brings the average way up. In this case, $310,000 is probably more representative of the home values in the area, but *both* values are true statistics.

In most geographic areas, a few homes are worth significantly more than the others; this usually makes the mean home value greater than the median home value. This trend also occurs with income and other financial markers. While the actual difference between the mean and the median in a real geographic area is usually less severe than it is in our example, there is usually a difference. By understanding the significance of that difference, we learn to see numbers for what they really tell us.

How can we be deceived by the use of these data? Consider this: if a newspaper article wishes to portray an areas as wealthy, which value would the writers use? Probably the mean since it is generally higher than the median. To portray the area as poorer, people often use the median. This may be more subtle than the graphical ones, but we can still be manipulated if we do not understand these terms and their use

Survey Deception

Some people enjoy taking surveys; others do not. I generally enjoy giving my opinion on things. However, what I frequently find more interesting is trying to figure out how I am being manipulated by the survey questions. While there are a number of classic tricks people use on surveys, people often simply write really poor survey questions, which skew the data. These people do not knowingly mislead us, but their results do not accurately reflect the truth they were seeking.

Here is a great example: I once gave a ten question survey about Kenya to two groups of students. The questions on the surveys were identical except for question 5. Question 6 on both surveys was "Estimate the population of Kenya." This question is innocent by itself. However, on the surveys for group A, question 5 read "Do you think the population of Kenya is greater or less than 20 million?" Group B's question 5 was "Do you think the population of Kenya is greater or less than 100 million?" How do you think the answers on question 6, "Estimate the population of Kenya," varied between groups A and B? As you might suspect, the average value for question 6 from group A was about 20 million; the average value from group B was about 100 million. Honestly, I think few people actually know anything about the size of Kenya's population. However, I conjecture that the shame of seeming uninformed caused people to use the seed value in question 5 to center their uninformed guess.

Let's look at a few categories of poor survey questions: Leading Questions, Double-Barreled Questions, and Double-Negatives. Each of these types has different levels of severity. While we study these, please remember that sometimes these questions are intentionally used for deception, but many times poor questions are just the result of inexperienced people writing surveys.

Leading Questions.

<u>Voting Question:</u>

"Would you vote for Bill Smith, a man known to accept questionable campaign contributions? (YES/NO/UNSURE)."

Clearly, this is a blatantly leading question because it loads any survey taker to be significantly negatively disposed to Bill Smith. Other leading questions are slightly more subtle. Look at these two possible ways to ask the same question.

<u>School Safety Bill Question:</u>

(1) "Do you support the School Safety Bill, a bill that will coordinate local and federal law enforcement and provide funding to place important security checks at the entrance to every school, ensure

sufficient police presence in the most dangerous schools, and provide professional psychological intervention in dangerous and difficult student situations? (YES/NO/UNSURE)."

(2) "Do you support the School Safety Bill, a bill that will unnecessarily divert millions of dollars from hiring the best teachers, reducing class size, reducing health care costs to millions, and providing economic stimulus to the region? (YES/NO/UNSURE)"

The same essential question is asked two different ways, leading those who are polled toward a particular answer.

The following question, although more neutral, *expects* a particular answer.

Emergency Medical Technician Question:
"Don't you agree that Emergency Medical Technicians (EMT's) should receive more pay for their first-responder service? (YES/NO/UNSURE)."

The phrase "Don't you agree" expects a positive answer. A more neutral question might me worded like this:

"Do you believe EMT's pay is higher than it should be, lower than it should be, or just about right?"

This last question's wording encourages people to make up their own mind, instead of encouraging a particular answer.

Double-Barreled Questions. I have found double-barreled questions to be the most frequent and most innocent type of problematic question on surveys, questionnaires, and presentation evaluations. In the noble effort to try to shorten a form to a reasonable length while still trying to gather as much data as possible, survey writers merge two questions into one. Try to answer these questions using the 5-point Likert scale (you know, "Strongly Agree, Agree, Neutral, Disagree, Strongly Disagree"):

1. "The presenter was well-prepared and used time wisely."
2. "The nurses attended to my needs quickly and pleasantly."
3. "The president is doing a good job handling domestic and foreign issues."
4. "University professors should be more involved with university admissions and marketing."

Do you see the potential problems with these survey items? Each item actually seeks an opinion on two different things. When considering item 1, I might think the presenter ended up using time wisely but did not have enough materials prepared for the presentation. In item 2, my nurse might have been grouchy by attentive. In item 3, I may think the president is a domestic policy whiz and a foreign policy buffoon. In item 4, maybe I think that university professors should help with admissions by contacting prospective students but are absolutely clueless on marketing. How should I respond to each item? The results of double-barreled questions are relatively useless because we do not know what is being measured. This practice is usually not a "deceptive" practice, but the results are assuredly misleading.

Double-Negative Items. Double-negative items include those classic true/false items that confuse the test-taker so that the assessment does not reflect what the test-taker knows. Although a survey participant may have a clear opinion or a test-taker may know an item, double-negative items confuse.

1. "Do you disagree that teachers should not do recess duty?
2. "True or false: George Washington was not the first president of the United States."

3. "Your doctor is not inattentive. (Agree/disagree)"
4. "The instructor should not schedule a paper to be due on the same day as an exam."

Items 1 and 3 both contain double-negatives, "disagree...not" and "not inattentive," which naturally create confusion. Items 2 and 4 are problematic because answering "no" or "false" creates a positive. Unless the test-giver or survey's goal is to confuse the participant—think of the ethical implications—these items should be avoided.

Something to Think About

Proverbs 12:22 says, "Lying lips are abomination to the Lord: but they that deal truly are his delight" (KJV), and Proverbs 11:1 says, "A false balance is abomination to the Lord: but a just weight is his delight." (A "balance" is another word for scale, by the way). Clearly, God tells us lying and deceiving are unacceptable. In the context of statistics, lies and deceptions are harder to identify. Intent becomes a big factor we must consider. We should also examine the extent of the "dishonesty." Since statistical deception comes in many forms, ethical questions become much more complicated. Consider all of the types of deceptions described in this chapter. Is every type equally problematic? Are all of the methods described in this chapter lies? Is it ever appropriate to employ the methods described in this chapter?

Covering the Reading

1. Which of the three graphical tricks described at the start of this chapter (Scale, Area, Position) do you find to be the most influential? Why?

2. In your own words, explain how the population density trick is deceptive.

3. Consider the ages of professional basketball players. Would you expect the mean age or the median age of professional basketball players to be higher? Why?

4. Re-write the "Voting Question" to improve its wording.

5. Write a question that would improve the "School Safety Bill Question."

6. Repair one of the double-barreled questions presented in this chapter.

7. Repair one of the double-negative questions presented in this chapter.

Problems

8. Using the following data set, develop a misleading graphical presentation. You can decide what the data are. Label the graph and write a "troubling," dramatic headline.

Republicans	YES	YES	NO	YES	YES	NO	NO	YES	YES	NO
Democrats	NO	YES	YES	YES	NO	YES	YES	NO	YES	YES

9. Write a survey containing at least 6 items that investigates people's attitudes and practices related to New Year's Eve. In the survey, use all of the poor techniques described in this chapter: Leading Questions, Double-Barreled Questions, and Double Negative Questions.

10. Locate a news article that reports either the mean or median for a set of people. How would the article be different if it used

the other statistic? Did the choice of mean or median impact the message of the article?

11. Is the "Scale Trick" described in this chapter a dishonest practice? Explain.

12. Is every method described in this chapter equally ethically problematic? Are all of the methods described in this chapter lies? Is it ever appropriate to employ the methods described in this chapter?

Chapter 2.5
Voting

Voting is one of the foundations of democracy. The ability to determine the will of a large group is central to the idea of self-governance. As we have all experienced on both small and large scales, it is relatively easy to determine the will of the people if they are given two options. One simply needs to find a way to determine which of the two options someone prefers. However, when there are more than two options, this task becomes considerably more difficult.

Warm-up Activity

In the 2010 Florida race for the U.S. Senate, Marco Rubio (R) received 43% of the popular vote, Charlie Crist (I) received 32% of the popular vote, and Kendrick Meek (D) received 20% of the popular vote. Obviously, in the United States voting system, the candidate who receives the most (first place) votes, wins the election.

Rubio	43%
Crist	32%
Meek	20%

But suppose we knew not only each individual's first choice, but also their second and third choices. If you prefer Crist over Rubio over Meek, then we could list your preferences as C-R-M.

This system allows for six possible preference orders: C-R-M, C-M-R, M-R-C, M-C-R, R-M-C, and R-C-M. The following data are hypothetical, but consider the implications of these preferences. Suppose 43% of the voters preferred Rubio over Meek over Crist (R-M-C), 20% had the preference of Meek over Crist over Rubio (MCR), 32% preferred Crist over Meek over Rubio, and the other three possibilities were not represented in the population.

43 %	20%	32%
R	M	C
M	C	M
C	R	R

Based on a popular vote or *plurality*, Rubio wins. However, consider the preferences carefully. Could you argue that the will of the people really chose a different candidate?

Concept Development
The world has developed many different voting schemes. In this chapter, we will consider six of them.

1. Plurality vote. In a plurality vote, the individual with the most first-place votes wins. In our Warm-up exercise, Rubio wins because of the 43% un-split vote, even though 52% of the people ranked him least favorite. This issue is a very common weakness of plurality voting; sometimes, the least liked candidate wins. **Winner: Rubio.**

2. Least favorite vote. This voting system has several different names, but the general idea is the same. It goes like this: determine which candidate is dead last on most people's preference list, and then eliminate that person from the race. In our Warm-up, the fact that Rubio is least liked removes him from the race; he cannot win.

Without Rubio in the race, Meek wins the race. We can determine this because we know people's preferences. Meek receives the 20% of the population that favors him most and then also grabs the 43% of the people who, with Rubio out, would vote for Meek. In the case that there are more than 3 candidates, you simply determine who is liked the least and eliminate that person. Then, you determine who is liked the least of the remaining people, and eliminate that person. Continue the process until only one candidate is left. **Winner: Meek.**

3. Run-off election. Many countries around the world use a run-off system. In this system, everyone votes for a candidate as you would in a plurality vote. Then the votes are tallied. A second round of elections then allows people to choose between the top two candidates. In our Warm-up, Crist and Rubio were the top two candidates in the first round of the election. Meek, and any other candidates, would be left out of the next round. People return to the polls and choose between Rubio and Crist. Based on our preference chart, Crist defeats Rubio in the run-off 52% to 43%. **Winner: Crist.**

4. Condorcet. The Condorcet method is a one-on-one method. Each possible pair of candidates is matched up, and we look for the king of the match-ups. The trouble with this method is that it frequently produces no winner. For example, A might defeat B, B defeats C, yet C defeats A: no clear winner! Our example case gives us three possible match-ups: Rubio vs. Crist, Crist vs. Meek, and Meek vs. Rubio. Using our preference chart, we know Crist would win the Rubio vs. Crist battle. Meek would win the Crist vs. Meek battle, and Meek would also win the Meek vs. Rubio pairing. Consequently, Crist is better than Rubio, but Meek is better than Rubio or Crist. **Winner: Meek**.

5. Borda Count. In the Borda Count voting method, each place (1st, 2nd, 3rd, etc.) is assigned a point value. All voters rate their preferences from first to last, and points are assigned accordingly. Since our example has three candidates, we will award 3 points for every first place vote, 2 points for every second place vote, and 1 point for every third place vote. If we had five candidates, we would assign 5, 4, 3, 2, and 1 points for respective preferences. Since we do not know exactly how many people voted in our example election, we will assume there were 95 voters representing the 95% total (I trust you noticed the percentages did not add up to 100 %). To do a pure Borda Count, we would need to know the exact number of voters and their preferences. Consequently, Borda Counts work well for small groups, but not well for larger ones. Let's calculate the Borda Count for each candidate:

Rubio received 43 first place votes (3 points each), no second place votes (2 points each), and 52 third place votes (1 point each). $(43 \times 3) + (0 \times 2) + (52 \times 1) = 181$ *points*.

Crist received 32 first place votes (3 points each), 20 second place votes (2 points each), and 43 third place votes (1 point each). $(32 \times 3) + (20 \times 2) + (43 \times 1) = 179$ *points*.

Meek received 20 first place votes (3 points each), 55 second place votes (2 points each), and no third place votes (1 point each). $(20 \times 3) + (55 \times 2) + (0 \times 1) = 170$ *points*.
Winner: Rubio

6. Elimination. In a three-candidate election, the elimination method is identical to the run-off method. However, for more than three candidates, it is basically a progressive run-off. Voters cast first-place votes. The candidate with the fewest first-place votes is eliminated. Voters vote again. The candidate with the fewest first place votes is eliminated. Voters vote again. The candidate with the fewest first place votes is eliminated. Voters vote again. The candidate with the fewest first place votes is eliminated. This process continues until all but one candidate is eliminated. If we have a preference chart (as needed for the Borda Count) then we can determine who the voters would progressively choose as candidates are eliminated. In the case above, Meek is eliminated in the first round with only 20% of the votes. With Meek eliminated, all of those voters turn to Crist (their second choice). Rubio is eliminated in round 2. **Winner: Crist.**

Something to Think About

Which of the three candidates do voters actually desire to elect? In a plurality vote and a Borda Count, Rubio wins. In a Condorcet election and Least-favorite vote, Meek wins. In a run-off and an elimination vote, Crist wins. That question is difficult to answer. Each voting system has its own strengths and weaknesses. Which one do you prefer the most? Which one should a country use to determine the will of the people? Which voting system should your country use? Maybe we should vote on it…

How can we accurately determine the will of the people? How can we make democracy work? Economist Kenneth Arrow, who won the Nobel Prize in 1972, demonstrated through Arrow's Impossibility Theorem that when there are three or more options, no voting system can ever be "consistent" and "fair." Arrow's theorem makes a few reasonable assumptions regarding voting systems such as each person getting only one vote. This theorem's impact remains significant: there is simply no completely fair way to determine the will of the people in a situation with three or more options. Every voting system has a weakness.

Covering the Reading

1. Choose two of the voting methods described in the chapter. Identify one strength and weakness of each.

For questions 2-7, use the preference chart below.
50 voters are choosing from among 4 candidates; their preferences are listed below. The first column shows that 14 people prefer candidate A over B over C over D. The second column shows that 13 people prefer candidate B over C over D over A. The third column demonstrates that 5 people prefer candidate C over B over D over A. The fourth column demonstrates that 7 people prefer candidate C over D over B over A. The fifth column demonstrates that 11 people prefer candidate D over C over B over A.

14	13	5	7	11
A	B	C	C	D
B	C	B	D	C
C	D	D	B	B
D	A	A	A	A

2. Who is the winner in a plurality vote?

3. Who is the winner in a run-off election?

4. Who is the winner in a least-favorite election?

5. Is there a winner in the Condorcet method?

6. Who is the winner in a Borda Count?

7. Who is the winner in an elimination method election?

Problems

8. Research and describe two additional voting methods.

9. Which voting method do you prefer? Why?

10. As you learned above, Arrow's Impossibility Theorem demonstrates that when there are three or more options, no voting system can ever be "consistent" and "fair." Discuss the impact of Arrow's Theorem on your view of democracy.

11. In a democracy, or at least a republic, we believe we determine our path based on our votes. Since those who wrote the Florida Constitution and laws determined that the United States would use plurality voting in the Florida election described in this chapter, you could argue that Florida voters indirectly and unknowingly chose Rubio to win the election. The Bible also offers perspectives on God's role in raising up and tearing down leaders. Reflect on and write out a resolution between our responsibility and beliefs as voting citizens and God's handling of the nations. What role do we play in the process?

Chapter 2.6
Probability

Unlike most of your past experiences with probability, this chapter is not primarily about calculating probabilities. Instead, we are going to look into the history of probability, the assumptions of probability, and their impact on society today.

In this chapter, I will assume that you know how to calculate some simple probabilities. For example, the probability of rolling a "4" on a standard six-sided die (die is singular for dice) is $\frac{1}{6}$. There are six possible outcomes, and one way to have a *success*. If we were to draw a single card from a standard deck of 52 cards, then the probability of drawing a "queen" is $\frac{4}{52}$. Probabilities are usually expressed as a fraction or a percent. For the purposes of improving our understanding, in this chapter we will use fractions that are not reduced. For example, if you were to roll a 6-sided die ("die" is the singular of dice), the probability of rolling an even number is three out of six. We will express that probability as $\frac{3}{6}$ instead of $\frac{1}{2}$ or 0.5.

We also need to quickly review a few other concepts as we begin this chapter. First is the difference between *theoretic probability* and *relative frequency*. If I were to flip a coin 200 times, and it came up heads 80 times, the relative frequency of heads would be $\frac{80}{200}$. Notice that relative frequency has to do with an experiment. However, if the coin were a fair coin then there are two possibilities: heads or tails. The theoretic probability of flipping a coin and getting heads is $\frac{1}{2}$. When people talk about the chances of being struck by lightning, they are not talking about the theoretic probability. The theoretic probability of being struck by lightning is very difficult to determine. However, it is easy to determine the relative frequency of being struck by lightning. We simply need to know the number of people in the world and the number of people struck by lightning. When we speak, we talk about the chances of being struck by lightning and the chances of flipping a heads on a coin. However, as you can see these two ways of using "chances" are quite different.

Additionally, in the world of probability, "success" is neither good nor bad. A success simply has to do with getting the hypothesized result. For instance, when calculating the relative frequency of getting struck by lightning, a *success* would be any individual who was struck by lightning. When we are flipping a coin and considering probabilities or relative frequencies related to heads, then heads relates to success while tails relates to failure.

You may recall a few more complexities from previous experiences with probability. Although this chapter is not primarily about calculating probabilities, it is important to recall some of the nuances of the subject. Consider the question of drawing a random card from the standard deck of 52 cards. What is the probability of drawing a queen? Answer: $\frac{4}{52}$. What is the probability of drawing a queen or a nine? Answer: $\frac{4}{52} + \frac{4}{52} = \frac{8}{52}$. However, suppose we wanted to know the probability of drawing a queen or a diamond. The answer is *not* $\frac{4}{52} + \frac{13}{52} = \frac{17}{52}$. What is wrong with this answer? There are 4 queens and there are 13 diamonds. However, we have counted the queen of diamonds twice: . The queen of diamonds was counted once when we counted the queens and once when we counted the diamonds. There are rules for handling instances when two events are *not mutually exclusive* ($P(A\ or\ B) = P(A) + P(B) - P(A\ and\ B)$). The point is that we must consider that probability is not always as simple as putting things together.

Secondly, consider the classic example of a bag containing 3 red and 5 blue marbles. What is the probability of randomly drawing a red marble? That is easy: $\frac{3}{8}$. However, what is the probability of drawing a red marble, replacing it into the bag, then drawing a red again? The solution for drawing a red, replacing, and drawing a red would be $\frac{3}{8} \times \frac{3}{8} = \frac{9}{64}$. Why do we multiply? Think of this idea in the context of rolling two dice. You can obtain a 1, 2, 3, 4, 5, or 6 on the first die, and a 1, 2, 3, 4, 5, or 6 on the second die. Each first die option (1, 2, 3, 4, 5, 6) is paired up with each second die option: 1 with 1, 2, 3, 4, 5, 6 and 2 with 1, 2, 3, 4, 5, 6, etc. $(6 + 6 + 6 + 6 + 6 + 6) = 6 \times 6 = 36$. And remember, repeated addition is the essence of multiplication. When we pair up, we have to multiply. Returning to our example of a bag of marbles, we need to consider one more complexity. What is the probability of drawing a red marble, *not* replacing it into the bag, and drawing another red marble? On the first draw, the probability of success is again $\frac{3}{8}$. However, if we assume success on the first draw and do *not* replace the marble into the bag, there would be only 2 red marbles left and 7 total marbles remaining: $\frac{2}{7}$. Therefore, the final probability is $\frac{3}{8} \times \frac{2}{7} = \frac{6}{56}$.

Whenever we calculate a probability, we really only need to know two things: the total number of possibilities and the total number of ways to succeed. These are two very elementary concepts; yet calculating probabilities can become quite complex. It would be nice if we could depend on our intuitions in the field of probability. Unfortunately, most people, including most mathematicians, have very poor intuitions when it comes to probability. A mathematical researcher named Efraim Fischbein conducted many studies on people's intuitions as related to probability. He repeatedly found that, unlike other areas of mathematics, our human intuitions regarding probability are unreliable. The famous Birthday Problem, for instance, simply asks how many random people need to enter a room before it is more likely than less likely that two of them share the same birthday (not including year). To put it another way, in a room of 30 people, would you expect two people to have the same birthday? In a room of 40 people? In a room of 50 people? How many people would it take to make the probability pass 50%? Would it take 100 people? 183 people? The answer may surprise you.

    I have one final comment to make before we proceed to the focus of this chapter. I have casually thrown the words "random" and "randomly" into the introduction, but these words should not be so casually applied. Randomness is at the heart of probability and, as we will see, statistics. This chapter will help us consider if there is such a thing as randomness.

Warm-up Activity

    To begin to understand the history and development of the study of probability, consider the following scenario:

    Suppose two people, Blaise and Pierre, agree to play a game involving repeatedly flipping a coin. Blaise calls "heads" and Pierre chooses "tails." The winner is the first one whose choice, heads or tails, appears 5 times. They each agree to pay $0.50 into a pot, and the winner keeps the $1.00.

    After they have flipped the coin 7 times, Blaise leads with 4 heads to Pierre's 3 tails. For an unknown reason, they have to quit the game at this point. How much of the pot should each person receive? Make an argument based on the probability of each person winning. Be sure your solution is *fair*.

Historical Context

In 1654, the gambler Chevalier de Méré proposed a problem to two mathematicians of the time: Blaise Pascal and Pierre de Fermat. Pascal and Fermat began to correspond with each other by letter. They were analyzing the "problem of the unfinished game" that had been circulating for a couple centuries but was reintroduced by de Méré. Although more complex than our scenario above, the unfinished game involved the division of stakes among gamblers for a game that was not completed. Through the pursuit of this problem, Pascal and Fermat made significant progress on establishing the field known today as probability. Pascal, Fermat, and many later mathematicians developed the mathematical rules for theoretic probability that we use today. They sought to be able to understand what had never occurred; they wanted to predict. This goal of early probability continued into the product of probability study: inferential statistics.

Concept Development

An important concept in the relationship between the natural, experimental world of relative frequency and the world of theoretic probability is the Law of Large Numbers. Consider the following scenario related to the law of large numbers. Suppose you were to flip a coin 10 times. You should not be surprised if heads comes up 7 times out of the 10 flips. However, if you flipped a coin 10,000 times and heads came up 7,000 times you should suspect that something is really fishy about that coin. The law of large numbers indicates that as the number of experiments increases, the relative frequency will approach the theoretic probability. Although $\frac{7}{10} = 70\%$ and $\frac{7000}{10000} = 70\%$, the difference between the theoretic probability of

50% and the relative frequency should be converging as the number of experiments grows.

The law of large numbers leads mathematicians to another observation. Suppose you were to roll a six-sided die 30 times. The results might not be perfectly even among the 6 options. However, as the number of rolls increases to 30,000, the distribution will look almost completely even or *uniform*. It is important to note that the experimental, relative frequency results match the theoretic probability of the random roll of a die. Probabilistic Theory matches real world behavior.

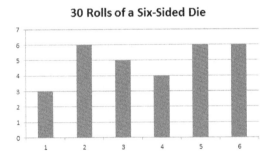

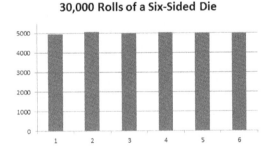

A uniform distribution is only one type of probability distribution; other probability distributions include binomial distribution, Poisson distribution, and normal distribution. Each distribution can be described theoretically with probability yet still models real world behavior. Consider a wait-time distribution for a subway line that arrives every hour on the hour. If people did not know the scheduled arrival time, then we would expect them to arrive at the station in a uniform distribution; people would arrive equally in each five-minute segment of the hour. However, if people are aware of the scheduled arrival time, then the distribution will be different. A few people will arrive very early. Some people will arrive 10 to 20 minutes early. Many people will arrive 5 to 10 minutes early, and many will arrive within 5 minutes of the scheduled time.

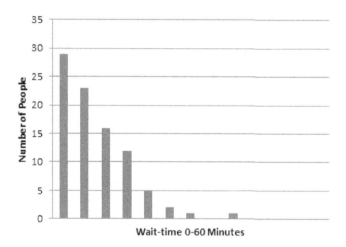

The normal distribution or "Bell Curve" is the most familiar distribution in popular culture. It is, as expected, in the shape of a bell and describes many natural phenomena such as heights of people: a few people are short, most are average, and a few people are tall. The normal distribution is a probability distribution based on a continuous function instead of discrete values. That means that you can calculate probabilities of decimal and fractional values within the distribution.

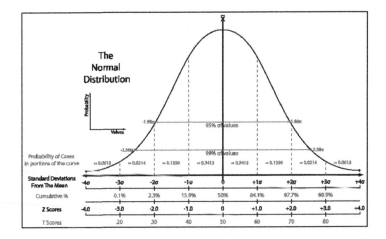

Another important component of statistics is the Central Limit Theorem. This theorem basically states that, given certain non-restricting conditions, the means of *any* distribution can be analyzed using the normal curve. In other words, regardless of how a certain distribution is actually modeled, there is a way to understand it through the normal curve given a large enough sample. This approach to modeling data with the normal curve forms the basis for today's inferential statistics, like the statistics used in experimental studies to draw conclusions about the population as a whole.

It is not an overstatement to say that probability is the basis for our study of the world today. Inferential statistics relies on the normal curve, and the normal probability distribution finds its basis in the study of probability. Medical, social, psychological, and business research all rely on the mathematics of inferential statistics. Think of it like this:

```
┌─────────────────────┐
│    Probability      │
└─────────────────────┘

┌─────────────────────┐
│ Normal Distribution │
└─────────────────────┘

┌─────────────────────┐
│ Inferential Statistics │
└─────────────────────┘
           ▼
┌─────────────────────┐
│    The Sciences     │
└─────────────────────┘
```

Something to Think About

Given that the study of probability finds its birth in wagers and games of chance, we should consider the wisdom and ethics of gambling. This path is a worthy of consideration. However, we should consider two more fundamental topics and two grand questions about the nature of the universe.

Much natural, behavioral, and social science research is based on the application of probability distributions through the use of inferential statistics in quantitative research designs. Research uses a relatively small sample to make generalizations to the entire population. At the heart of this practice is the belief that the universe is consistent and reliable. In other words, the height distribution of 1,000 people is the same as that of 1,000,000 people. Is there a biblical basis for this belief? On what basis can we trust that research conducted on a group will apply to a larger population?

As mentioned earlier in the chapter, theoretic probability is based on a concept of randomness. Inferential statistics research requires its researchers to be able to generate a random sample. Randomness in this context does not mean chaotic or scattered; instead, when we talk about a random selection, we mean that every option has an equal chance of being chosen. There is no favoritism or guiding hand. However, we know from the Old Testament that God clearly indicated his will by the drawing of lots or the use of the Urim and Thummim. God indicated Saul would be king by lot. David regularly received the will of God through the high priest's use of the Urim and Thummim. Judas Iscariot's replacement was chosen by lot. God can and has used "random" objects to communicate his will.

Now let's back up and ask two more global questions. First, to what extent does the will of God extend to the rolling of dice? Does God control every roll? If I flip a coin and it comes up heads, did God decide it would come up heads, or did He let it happen? The second question relates to determinism and the reliability of the universe we discussed before. Is anything truly random in the universe or is everything determined by the existing conditions? In other words, if we have two situations in the universe in precisely the same set-up, will the outcome be exactly the same? Could two identical situations have two different outcomes?

Covering the Reading

1 – 4. Suppose a single card is drawn from a standard deck of 52 cards. On a single draw, what is the probability of drawing:

1. The ace of spades?

2. A face card (jack, queen, king)?

3. A heart or diamond?

4. A face card, heart, or diamond?

5. Summarize the difference between theoretic probability and relative frequency.

6 – 7. Sketch the shape of the distribution you predict for each situation:

6. Shoe sizes of adult women.

7. Time spent using technology by different age groups (broken up into 10 year increments).

Problems

8. Locate a version of Fischbein's probability quiz. Take the quiz and report your results. Which questions revealed your poorest intuitions?

9. Is there a biblical basis for the consistency and reliability of the universe? On what basis can we trust the application of research to a larger population?

10. To what extent does the sovereignty of God extend to the rolling of dice? Does God control every roll? If I flip a coin and it comes up heads, did God decide it would come up, or did He let it happen?

11. Is there anything that is truly random in the universe or is everything determined by the existing conditions? In other words, given two situations in universe that are in precisely the same set-up, will the outcome be exactly the same? Could two identical situations have two different outcomes?

Chapter 2.7
Complexity and Chaos

Complexity theory and *chaos theory* are related but separate fields of study. Most of this chapter will be spent considering the commonalities of the two: sensitivity to initial conditions. However, we will return to their differences as the close of the chapter.

Warm-up Activity

Initial Conditions. We need to explore two ideas as we engage this chapter. The first concept is that of small differences in *initial conditions*. The second concept is *complex systems*. Consider this simple situation. Picture a cue ball on a pool table. What happens to the trajectory of the cue ball when struck at *slightly* different angles? Suppose we strike the ball with a cue stick, with a lot of force, but we put no spin on the ball. We are now going to project the path of the ball over the first 6 times it strikes a bumper, the sides of the table. Remember that the *angle of incidence* equals the *angle of reflection*; in other words, the incoming and outgoing angles to and from the bumpers are identical. If you have one, use a protractor for this activity. Carefully project the path of the ball if it begins on *Path A* and if it begins on *Path B*.

The initial angle and distance between the two paths is very small: a few degrees and a few inches. How would you describe the difference between the two paths after the ball strikes the bumper 6 times? Although there was only a slight difference in *initial* trajectory, was the *slight* difference maintained or was the position of the two balls after the six ricochets significantly different?

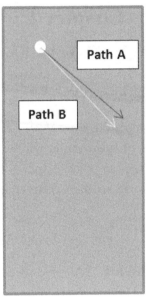

Complex Systems. One obviously complex and difficult system to predict is the weather. Most forecasts lack accuracy because so many factors affect the weather. Can you list five other systems that are complex in the same sense as the weather?
1.
2.
3.
4.
5.

Historical Roots

James Gleick's 1987 book *Chaos: Making a New Science* launched Chaos Theory into the public consciousness. The mathematical roots behind the theory relate to Benoît Mendelbrot's work on fractals, but the theoretical developments in this field are attributed to Edward Lorenz, the man who coined the famous term "The Butterfly Effect."

The Butterfly Effect is the theoretical story of a butterfly in one part of the world that induces a tornado in another area of the world. The main idea is that if one butterfly flaps its wings, it causes a slight stir in the air, which causes just the right twist in the atmosphere, that allows just the right systems to join together, that result in a tornado in Kansas. If the butterfly had not flapped its wings at that moment, the chain of events would not have begun, and the tornado would have been averted.

Concept Development

The Butterfly Effect illustrates a few concepts essential to Complexity Theory and Chaos Theory. Contrary to its title, the book *Chaos* does *not* present our world as chaotic. Instead, it explains how the world contains systems so complex that their predictability is extremely challenging, even impossible. The complexity of a system makes it difficult to predict exactly what will occur. Nonetheless, we can usually predict what will *not* occur. Every flap of a butterfly's wing does not produce a tornado. We can say with confidence that in general, butterfly-wing-flapping will not produce tornados. But predicting *exactly what will happen* is extremely difficult.

Complex systems are potentially quite sensitive to initial conditions. Slight differences in starting points *could* result in profoundly different chains of events. However, they may not. This concept led Lorenz to summarize Chaos Theory like this: "Chaos: When the present determines the future, but the approximate present does not approximately determine the future." We can illustrate the concept this way: suppose I dropped two small sticks from a bridge into a stream. The sticks are approximately the same size and weight, and I dropped them into approximately the same place in the stream at approximately the same time. Yet due to the complex nature of fluid dynamics, those two sticks could end up in miles apart and in two completely different conditions in just a few days. However, they may not.

Applications

This concept is best studied through a few well-chosen examples. In each case, we will see three recurring features:
1. **Myriad Variables:** We will be able to identify many factors that feed into the system, making the outcomes difficulty to determine.
2. **Predictability:** We can anticipate what will not happen with strong certainty, but will have difficulty anticipating exactly what will happen.
3. **The Butterfly:** We will be able to identify something seemingly minor that *might* have a major impact. Remember, complex systems are sensitive to initial conditions.

The Weather. One of the most familiar examples of a complex system that is difficult to predict is the weather . We have all experienced the forecasted "storm of the century" that actually whimpered instead of roared. This past winter in the mid-Atlantic region, we also experienced a predicted "light dusting" of snow that turned out to be six inches deep. Why is weather forecasting so frequently and consistently wrong? Because of the complexity of the system. You have probably noticed a few features about weather forecasting. For example, the accuracy of prediction 5 weeks out is borderline guessing while the accuracy 5 days out is sometimes satisfactory, and the accuracy of prediction 5 hours out of a certain time is almost always correct. Prediction becomes easier closer to the event.

Furthermore, consider our three categories:
1. Myriad Variables: What kinds of things can impact local weather forecasting? If you visit the National Oceanic & Atmospheric Administration (NOAA) webpage and search for the datasets used to predict weather, you will find dozens of

categories and sub-categories of data used to attempt to predict the weather.

2. Predictability: When weather forecasters post a 5-day forecast, they are usually wrong about the temperature by a few degrees, wrong about the exact timing and amount of any precipitation, and vague about the amount of cloud cover. It is very difficult to predict exactly what will happen. However, they can be fairly confident about what will not happen. If it appears that a major snowstorm is heading for a region, their exact predictions will be off. However, they can be highly confident that it will not be sunny and 78 degrees. Likewise, if they are debating between partly cloudy and mostly cloudy, they can feel certain that it is not going to rain six inches.

3. The Butterfly: The Butterfly Effect is so named because of its potential impact on the weather. Can you think of a more tangible example? How could turning a few fields into parking lots impact the weather? Aside from the geological impact involving rain run-off, even a small amount of urbanization alters the local temperature by a small amount. Cities are generally a few degrees warmer than the surrounding rural areas. Those few degrees can make the difference between rain, ice, sleet, and snow. Could a few wind turbines or a skyscraper impact the flow of air enough to complicate local weather forecasting? Could the building of a man-made lake produce enough evaporation to turn what would have been a small snowstorm into a major event?

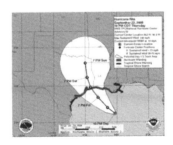

Microloans and gifts. Economists have recognized that a relatively small amount of money used in a strategic place can have a major impact on an individual, tribe, community, clan, or culture. A microloan ranges from a $25 loan to an individual in a developing country to a $50,000 for a small business entrepreneur in a developed country. Similarly, some organizations orchestrate the purchase and delivery of small gifts that could make a significant difference in a family's livelihood. Organizations help you or me to purchase a few chickens for a family living in poverty. The hope is that the eggs will both help feed the family and possibly allow them to generate a little income. In turn, they may be able to increase their flock, thereby multiplying their income and pulling themselves out of poverty. The possibilities are endless. On the other hand, instead of changing a family's life, the chickens could die.

1. Myriad Variables: Suppose you sent a small (for you) loan of $25 or a gift of two milk-producing goats to an impoverished farmer. Can you think of many factors that could produce significantly varying results?

2. Predictability: It is difficult to predict the exact impact of microloans and gifts. There could be little to no impact, or a family's life could be dramatically altered. However, in the large majority of cases we can say with confidence that the receiver will not become a millionaire or billionaire. We can also say that the gift will likely not result in the person's death. However, it is very difficult to say exactly what will happen; we can only predict what could happen with different levels of confidence.

3. The Butterfly: The microloan and gift practices are in a sense dependent on the Butterfly Effect. The hope is that a small, seemingly minor event can result in a major impact.

Our lives. Malcolm Gladwell's book, *Outliers: The Story of Success*, features a number of stories about "successful" people.

Throughout the book, he contrasts people who had very similar beginnings and natural talents but walked different paths in life. The essence of his thesis is that those who take advantage of certain specific, key opportunities diverge from those who do not. These opportunities often seem small and insignificant or may be rare. In every case, though, society's complex system of life and development is significantly impacted by what appear to be relatively minor events.

1. Myriad Variables: I am sure you can think of thousands of factors that impact the paths of our lives. Parents, location, intellect, talent, etc.

2. Predictability: Apart from the promises of our belief system, life is quite unpredictable. This unpredictability is especially evident when we think about what is likely to happen 5 minutes, 5 days, 5 weeks, 5 years, or even 5 decades from now.

3. The Butterfly: You can probably identify "butterflies" in your past: events or opportunities that may happen every day to billions of people. However, in your case, the event has had significant, lasting impact on who you are and what you do. On the other hand, events and decisions we make every day frequently have no lasting impact, but they do have the potential to impact and morph the rest of our lives.

Other Examples. Books, movies, and even history offer us many other examples of the Butterfly Effect. Bilbo Baggins finds the One Ring. Anne, and not a boy, is sent to Green Gables. Peter Parker is bitten by a spider. Dorothy chases Toto. Neo takes the red pill. How any other examples can you think of? In each case, something relatively minor or mundane results in something significant.

Something to Think About

This chapter asks us to consider at least two significant implications of Chaos Theory. The first has to do with the nature of the universe. Is the universe deterministic? In other words, can everything that happens be explained purely as cause-and-effect? The second has to do with the impact of the choices we make every day. Let's examine each of these in turn.

Related to Chaos Theory, the question of determinism goes back to Sir Isaac Newton. His belief in God notwithstanding, most people agree that one impact of Newton's work in physics made God, in a sense, obsolete. The mathematics of Newtonian physics allows for this kind of certainty: if we know the exact starting conditions and circumstances of any physical, natural phenomenon, then we can predict with perfect accuracy what will happen in that particular case. This power over nature was very rewarding! Yet this ability came with a price. If the universe is perfectly predictable, then what occurs tomorrow is a direct result of the conditions today and cannot be changed. Tomorrow's events are already determined. If we work backward through time, the events of today are a direct result of the conditions of yesterday. This argument can then be traced all the way back to the dawn of time, and we are forced to conclude that who and what we are today are only and exactly the one possible result of history. We can also make this argument about the chemical and electrical events that occur in our brains. In this case, we could say that the decisions we make are not really decisions but are pre-determined by our exact physical condition at that moment.

How does Newtonian physics relate to complexity and chaos? Chaos Theory claims that systems are so complex that the ability to predict may elude us. However, a particular belief underlies Chaos Theory: that if we were able to understand all of the many, complex variables in any given situation, we could predict and determine its results. Complexity Theory diverges from Chaos Theory at this point. Complexity Theory allows for two identical situations to have two different results. This is a direct contrast to Chaos Theory's premise that the universe is deterministic. What do you believe about the nature of the universe?

In the early 2010's, an app called *Angry Birds* became wildly popular. In this game, different types of birds were propelled from a sling-shot in an effort to destroy the evil pigs. The pigs were protected by many configurations of fortresses which the birds had to penetrate. I do not know the programming code used for this game, but the number of possible ways to destroy the pigs with a series of birds is huge. In a sense, this simple game meets some of the criteria for a complex system. In some ways, the game is deterministic: if you did the exact same thing in the exact same way, then the exact same result would occur. However, to keep people interested, the code could have been written with a pseudo-random number generator that would cause the user to have potentially different results when doing exactly the same thing in exactly the same way. The question we might consider is "How is our universe like and unlike *Angry Birds*?"

A second line of question worth considering is the path of our own lives. The first stanza of Robert Frost's 1920 poem *The Road Not Taken* reads:

> Two roads diverged in a yellow wood,
> And sorry I could not travel both
> And be one traveler, long I stood
> And looked down one as far as I could
> To where it bent in the undergrowth…

The poet then debates which path he should take. He chooses a road, and he finds that that particular choice "has made all the difference."

Each day we face many choices. Most of them seem insignificant, but if our lives are complex systems, then perhaps the Butterfly Effect should come to bear on our thinking more often. What am I doing today, now, that may have a significant, lasting impact on my life? The path I choose today can make "all the difference" later.

Covering the Reading

1. Complete the Warm-Up exercises. In the case of the pool game, was the *slight* difference maintained? What are the complex systems that you identified?

2. Summarize "The Butterfly Effect."

3. Restate Lorenz's quote in your own words: "Chaos: When the present determines the future, but the approximate present does not approximately determine the future."

4. Identify a complex system *not* described in this chapter. Explain how it meets the three criteria described: Myriad Variables, Predictability, and The Butterfly.

Problems

5. Provide three examples of books or movies that illustrate the Butterfly Effect and sensitivity to initial conditions as described in this chapter. Be sure to identify the relatively small event and the relatively large consequence for each one.

6. Write two different bullet-point story outlines. Make the first three bullets identical. The fourth bullet should be slightly different; incorporate a minor, seemingly insignificant difference. Make the stories diverge significantly as a result of the fourth bullet point..

7. Identify two or three seemingly minor events in your personal history that have had lasting major impacts.

8. Reflect on the discussion of the deterministic nature of the universe as described in this chapter. Use the *Angry Birds* illustration if it helps you make your point. If the exact same thing was done twice, would the exact same results occur?

9. Identify and reflect on the potential impact of one minor, bad academic decision that you could make this school year. It may

be academic dishonesty, failing to study, or anything else related to school. What is the potential impact of that bad decision?

Part 3
Number

Chapter 3.1
Special Numbers

In this chapter, we will examine different types of numbers. I'm not talking about the differences among rational numbers, real numbers, imaginary numbers, whole numbers, etc. Instead we are only going to examine properties of the whole numbers, $\{0, 1, 2, 3, 4, \cdots\}$, and its subset, the natural numbers $\{1, 2, 3, 4, \cdots\}$. These numbers can be categorized in more ways than you can imagine. Let's investigate a few of these numbers' special properties to determine what makes specific numbers special.

$$\{1, 2, 3, 4, 5, ...\}$$

Warm-up Activity 1
Try to list the first 5 numbers of each type.
  A. Primes:
  B. Squares:
  C. Cubes:
  D. Triangular:
  E. The sum of the first n natural numbers:
  F. The product of the first n natural numbers:

Concept Development 1

These sets of numbers are fairly familiar or easily calculated numbers. 25 is a square number. 23 is a prime number. 24 is the product of the first 4 natural numbers. Each of these numbers has special properties. Throughout this chapter, we will question whether one number is better than another because of the types of properties or number of properties each number has. For example, among other properties, 120 is the product of the first 5 natural numbers, it is a triangular number, and it is $3^1 + 3^2 + 3^3 + 3^4$. Is 120 more special than 119 or 121? We will also pause to consider whether we like a number more than another for mathematical reasons and for personal reasons. For example, I like the numbers 1997 and 1999 because they are the most recent pair of years that are twin primes. I also liked being ages 33, 34, and 35 because they were a sequence of three years that were the product of two primes: $33 = 3 \times 11$, $34 = 2 \times 17$, and $35 = 5 \times 7$. As you can see, I have a bias toward prime numbers.

Warm-up Activity 2

These numbers may be less familiar. Try to list the first 5 numbers of each type.

    A. Fibonacci numbers:
    B. Lucas numbers:
    C. Perfect numbers:
    D. Abundant numbers:
    E. Deficient numbers:
    F. Friendly numbers:

Concept Development 2

The Fibonacci numbers and Lucas numbers are closely related. Both the Fibonacci numbers and Lucas numbers are the sum of the previous two numbers in the sequence. The difference lies with the starting numbers. Remember the Fibonacci Sequence? $0, 1, 1, 2, 3, 5, 8, 13, 21, \cdots$. Begin with 0 and 1 or begin with 1 and 1. Each number is the sum of the two previous numbers. For example, $13 = 5 + 8$. The Lucas numbers have the same property, except the sequence begins with 2 and 1.

Perfect, abundant, deficient, and friendly numbers also have similarities. When I ask people to list a perfect number, many answer 7. However, mathematically, 7 is deficient. Every natural number is either deficient, abundant, or perfect. For our purposes here, we are going to use a more informal definition for each of these. Consider the proper divisors of any number. For example, the proper divisors of 24 are 1, 2, 3, 4, 6, 8, and 12. If we add together the proper divisors of 24, the sum is $1 + 2 + 3 + 4 + 6 + 8 + 12 = 36$. Since $36 > 24$, 24 is an abundant number. However, if we add the proper divisors of 15, we find the sum is $1 + 3 + 5 = 9$. Since $9 < 15$, 15 is a deficient number. If the sum of the proper divisors of a number exceeds the number, then the number is considered *abundant*. If the sum of the proper divisors of a number is less than the number, then the number is considered *deficient*. What is a perfect number? If the sum of the proper divisors of a number equals the number, then the number is *perfect*. 6 is the first perfect number. $1 + 2 + 3 = 6$. The proper divisors of 496 are $1, 2, 4, 8, 16, 31, 62, 124, and\ 248$. 496 is also a perfect number since $1 + 2 + 4 + 8 + 16 + 31 + 62 + 124 + 248 = 496$.

Amicable numbers, also known as amicable or friendly pairs, apply the same concept as deficient, abundant, and perfect numbers with a twist. 220 and 284 form an amicable pair because the proper divisors of 220 add up to 284, and the proper divisors of 284 add up to 220. In other words, 220's abundance adds to 284 and 284's deficiency adds to 220.

Proper divisors of 220: $1, 2, 4, 5, 10, 11, 20, 22, 44, 55, 110$

Sum of the proper divisors of 220:

$1 + 2 + 4 + 5 + 10 + 11 + 20 + 22 + 44 + 55 + 110 = 284$

Proper divisors of 284: $1, 2, 4, 71, 142$

Sum of the proper divisors of 284:

$1 + 2 + 4 + 71 + 142 = 220$

Just this quick review of perfect, abundant, deficient, and amicable numbers, shows us how numbers can be classified based on playing with their divisors.

Warm-up Activity 3

You may have known about the numbers we just looked at, but the following sets of numbers are even less common in the K-12 curriculum. Instead of trying to list these numbers, make a guess as to what each set is.

A: Palindromic numbers:

B: Happy numbers:

C: Narcissistic numbers:

D: Powerful numbers:

E: Vampire numbers:

Concept Development 3

Palindromic numbers. If you know what a palindrome is, you were probably able to guess what palindromic numbers are. A palindrome is a word or phrase that is spelled the same backward and forward. For example, MOM, HANNAH, and RACECAR are all palindromes. Similarly, 167761, 8118, 273898372, and 22 are palindromic numbers. That one was fairly easy.

Happy numbers. I doubt you were able to guess any of the rest of these types of numbers. However, once you see what they are, their names should make more sense. In order to understand happy numbers, you need to understand a process.

1. Given a number,

For example: 91

2. Take the digits of the number. Square them and add them.
$$9^2 + 1^2 = 81 + 1 = 82$$

3. Repeat step 2 with the resultant number.
$$8^2 + 2^2 = 64 + 4 = 68$$

4. Repeat step 2 with the resultant number over and over again.
$$6^2 + 8^2 = 36 + 64 = 100$$
$$1^2 + 0^2 + 0^2 = 1 + 0 + 0 = 1$$
$$1^2 = 1$$

One of two things will happen. Like 91, 82, 68, and 100, the repeated pattern will end with a 1. The other option is that the number sequence will go into a loop that never gets to 1. For example,

$$4$$
$$4^2 = 16$$
$$1^2 + 6^2 = 37$$
$$3^2 + 7^2 = 58$$
$$5^2 + 8^2 = 89$$
$$8^2 + 9^2 = 145$$
$$1^2 + 4^2 + 5^2 = 42$$
$$4^2 + 2^2 = 20$$
$$2^2 + 0^2 = 4$$

Which numbers do you think are happy? 91 is happy, but 4 is not happy. Consequently, 91, 82, 68, and 100 are also happy while 4, 16, 37, 58, 89, 145, 42, and 20 are not happy.

Narcissistic numbers. As you might suspect, just as a person who is narcissistic is overly focused on himself, narcissistic numbers are also excessively preoccupied with themselves. 153 is a 3-digit narcissistic number because $1^3 + 5^3 + 3^3 = 153$. Similarly, 8208 is a 4-digit narcissistic number because $8^4 + 2^4 + 0^4 + 8^4 = 8208$. Every 1-digit number is narcissistic: $1^1 = 1; 2^1 = 2; 3^1 = 3$, *etc.* 115,132,219,018,763,992,565,095,597,973,971,522,400 is also narcissistic.

Narcissus **by Caravaggio**

Powerful numbers. Powerful numbers have to do with powers or exponents. The prime factorization of any powerful number never has a power or exponent as small as 1. 72 is powerful because its prime factorization, $72 = 2^3 \times 3^2$, has no factors to a power of 1. However, 250 is not powerful because in its prime factorization, $250 = 2 \times 5^3$, 2 has a power of 1. More technically, a number n is powerful if, when p *evenly divides* n, then p^2 *also evenly divides* n. You cannot have an exponent or power of 1 in your factorization and be powerful.

Vampire numbers. A vampire number always has 2 fangs. First, a vampire number has an even number of digits such as 1435. We split the digits in half, in this case 2 digits and 2 digits, under the following conditions. If we can find *any* way to take two halves, or two fangs, such that the product of the fangs is the original number, then the number is a vampire number. So, if we can take the 4 digits of 1435 and split them into *any* two 2-digit numbers (fangs) that multiply to be 1435, then 1435 is a vampire number. It happens that $35 \times 41 = 1435$, so 1435 is a vampire number. 2187 is also a vampire number because we can form two fangs, 27 and 81, whose product is 2187. $27 \times 81 = 2187$. Some vampire numbers even have two different pairs of fangs. $12054060 = 2004 \times 6015$ and $12054060 = 2406 \times 5010$.

Historical Spotlight

A number of years ago while attending a mathematics conference, I had the privilege of witnessing Rev. Stanley Bezuszka, S.J., share his love for numbers. While in his 90's teaching at Boston College, he visited the conference and spoke on many beautiful and intriguing number patterns and types. His love for numbers was contagious! No one left the session without a deeper understanding and appreciation for the lure of special numbers. His books *Number Treasury* and *Number Treasury 2* have survived him after his passing in 2008.

Something to Think About

People often ask about the applications of perfect numbers or happy numbers or vampire numbers. As far as I can find, there is no application. We may find one in the future but, quite frankly, to many mathematicians, these numbers do not require applications to be wonderful. In his book *A*

Mathematician's Lament: How School Cheats Us Out of Our Most Fascinating and Imaginative Art Form, Paul Lockhart bemoans the loss of mathematics as a beautiful, enjoyable, and even recreational subject. Mathematicians like Lockhart regard the question of the applicability of perfect numbers as parallel to asking about the applicability of a great work of art. Stanley Bezuszka enjoyed numbers and their properties with or without a direct application. "When am I ever going to use this?" was not part of his lecture on special numbers. In a 2007 *Doctor Who* episode, the Doctor asks, "Don't they teach recreational mathematics anymore?" If many mathematicians consider mathematics apart from its application, should we do the same too? Should there be such a thing in our lives as recreational mathematics? Perhaps the most important question is, what is a biblical perspective on mathematics? Does it have to be applicable to be good? Can something that is simply enjoyable or even beautiful be good in its own right? Should the Christian be pursuing recreational mathematics?

The other question worth reflecting upon is "Are some numbers more special than others?" I suppose the question presumes that numbers are special. As we have seen, different numbers have different properties. These properties are characteristics of the number. Are some of those characteristics more important or more valuable than others? Are some numbers better than others?

Finally, is it okay to like some numbers more than others because of their mathematical characteristics? Maybe you are wondering why you would like a number for its mathematical characteristics in the first place. Do you enjoy a particular special number property?

Covering the Reading

1. From your list of squares and cubes, find two "squbes"-numbers that are both squares and cubes.

2. Determine and demonstrate if each of the following numbers is perfect, abundant, or deficient: 25, 26, 27, 28, 29.

3. Show that 1184 and 1210 are an amicable pair.

4. Determine and demonstrate if each of the following numbers is happy: 78 and 79.

5. Show that 9474 and 371 are narcissistic numbers.

6. Show that 1530 is a vampire number.

Problems

7. Based on your understanding gained from this chapter, define a happy prime. Avoid the temptation of looking it up and think about it. Then list one.

8. Define your own special number type based on a mathematical property, and list at least 5 numbers of that type.

9. Choose a number between 10 and 40, and find 3 mathematically special things about your number. They do not have to be "special" likethe properties studied in this chapter.

10. $135 = 1^1 + 3^2 + 5^3$. Verify that 518 and 598 have the same property. Can you find another one between 170 and 190?

11. $3435 = 3^3 + 4^4 + 3^3 + 5^5$. Verify that 438,579,088 has the same property.

12. Is mathematics important to study apart from its application? Should there be such a thing in our lives as recreational mathematics? Does it have to be applicable to be good or is something that is simply enjoyable or even beautiful good in its own right? Defend your response biblically.

13. Are some of the number characteristics we studies in this chapter more important or more valuable than others? Are some numbers better than others?

14. Do you enjoy some special number property? Explain.

Chapter 3.2
Prime Numbers

Mathematicians love prime numbers. As you recall, one way that the whole numbers greater than 1, {2, 3, 4, 5, 6, 7, ... }, can be divided into categories is prime versus composite. A prime number is a number greater than 1 whose only *divisors* are 1 and itself. Consequently, 5 is prime but 6 is not because $2 \times 3 = 6$. We say "2 and 3 both divide 6 evenly."

Warm-up Activity

Identify each of the items below. Explain what each one is. Then identify what the three have in common.

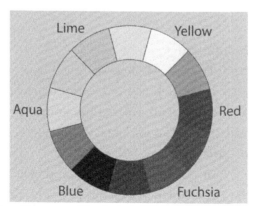

http://commons.wikimedia.org/wiki/File:Hsv_color_circle.svg

http://commons.wikimedia.org/wiki/File:Dodekafone.jpg

http://commons.wikimedia.org/wiki/File:Periodic_Table_Armtuk3.svg

Did you figure it out?

Concept Development

The periodic table of elements, the color wheel, and the 12-tone scale all serve as building blocks, or elemental objects, when molecules, colors, and musical harmonies are constructed. What is the mathematical equivalent? Prime numbers.

There are several "Fundamental Theorems" in mathematics. For example, Newton and Leibniz elegantly connected the Fundamental Theorem of Calculus.. The Fundamental Theorem of Algebra provides the basis for finding "roots" of polynomials. At a more foundational level, we find the Fundamental Theorem of Arithmetic.

The Fundamental Theorem of Arithmetic. The Fundamental Theorem of Arithmetic states that any whole number greater than 1 can be uniquely written as a product of prime numbers. Since this theorem includes many concepts and terms, let's consider them one at a time. First, a "whole number greater than 1" is the set $\{2, 3, 4, 5, 6, 7, \cdots\}$. Before we consider the "uniquely written" portion, examine the phrase "product of prime numbers." You likely recall that "product" means multiplication. We have already defined a prime number as a whole number greater than 1 whose only divisors are 1 and itself. Here is a helpful short list of primes:

$2, 3, 5, 7, 11, 13, 17, 19, 23, 29, 31, 37, 41, 43, 47, 53, 59, 61, \ldots$

When we say that each whole number greater than 1 can be written as a "unique" product of primes, we are saying that there is one and only one way to express a number as a product of primes.

Here are a few examples:
$$24 = 2 \times 2 \times 2 \times 3 = 2^3 \times 3$$
$$51 = 3 \times 17$$
$$550 = 2 \times 5 \times 5 \times 11 = 2 \times 5^2 \times 11$$

Each number has one and only one way to factor it into prime numbers. You may argue that $51 = 3 \times 17$ and $51 = 17 \times 3$ so the factorization is not unique. However, since multiplication has the *commutative* property, $3 \times 17 = 17 \times 3$, or three times seventeen is the same as seventeen times three. Consequently, since both expressions are the same, the factorization is unique. The key point is that we cannot break 51 down into different elements. It always breaks down to the same prime numbers: 3 and 17.

Prime numbers are foundational to the Fundamental Theorem of Arithmetic, which itself is foundational to number theory. Number theory is foundational to algebras and in turn to calculus and all sorts of mathematics unfamiliar to most people. In other words, prime numbers are foundational to mathematics.

At this point in their study, people frequently ask "Why isn't 1 prime?" The answer is both simple and frustrating. The number 1 is not prime because we define primes to exclude 1. Here is the problem: if we defined primes to include 1, then every theorem would need an extra line regarding 1. For example, if 1 were prime, then $51 = 3 \times 17 \times 1$ or $51 = 3 \times 17 \times 1 \times 1$ or $51 = 3 \times 17 \times 1 \times 1 \times 1$, and the factorization of 51 would not be unique. The Fundamental Theorem of Arithmetic would have to be rewritten to exclude 1: "Every whole number greater than 1 can be written uniquely as a product of primes if powers of 1 are not considered." Furthermore, every theorem involving primes would need an extra line that excludes 1 in order to be significant. The better solution by far is not to include 1 in the list of primes from the start. Do not feel bad for 1; it holds its own unique position as the multiplicative identity: it is the only number that has no impact when you multiply with it.

Finding Prime Numbers. How can we check if a number is prime or composite? Easy. Suppose we want to check the number 91. We could simply try to divide 91 by each prime to see if it divides evenly (leaving no remainder). 2 does not divide 91 evenly. 3 does not; nor does 5. 7 does divide 91 evenly. As long as we check the primes up to the square root of the number, we can see if it is prime or composite. (Why only to the square root of the number? If we find a prime factor greater than the square root, it will be paired with a prime less than the square root, which we would already have found in our process). The weakness of the approach I described is that we need to have a list of prime numbers to determine if a larger number is prime. More complicated modern methods use technology and some powerful theorems, but an old Greek method (again!) serves our purposes better here.

The Seive of Eratosthenes (276-194 BC) is a classic, brute force method for finding prime numbers. Its few simple steps are illustrated here. Suppose we want a list of all primes from 2-50.

Step 1: List the numbers:

	2	3	4	5	6	7	8	9	10
11	12	13	14	15	16	17	18	19	20
21	22	23	24	25	26	27	28	29	30
31	32	33	34	35	36	37	38	39	40
41	42	43	44	45	46	47	48	49	50

Step 2: Highlight or circle the first prime number 2 and eliminate all of its multiples since the multiple of a number cannot, by definition, be prime.

	2	3	~~4~~	5	~~6~~	7	~~8~~	9	~~10~~
11	~~12~~	13	~~14~~	15	~~16~~	17	~~18~~	19	~~20~~
21	~~22~~	23	~~24~~	25	~~26~~	27	~~28~~	29	~~30~~
31	~~32~~	33	~~34~~	35	~~36~~	37	~~38~~	39	~~40~~
41	~~42~~	43	~~44~~	45	~~46~~	47	~~48~~	49	~~50~~

Step 3: Highlight the next available number. It will be prime. Then eliminate all of its multiples. Some numbers might be repeatedly eliminated.

	2	3	4	5	6	7	8	9	10
11	12	13	14	15	16	17	18	19	20
21	22	23	24	25	26	27	28	29	30
31	32	33	34	35	36	37	38	39	40
41	42	43	44	45	46	47	48	49	50

Step 4: Repeat step 3 until you reach the square root of the highest number in your range (in this case about $\sqrt{50} \approx 7.1$).

	2	3	4	5	6	7	8	9	10
11	12	13	14	15	16	17	18	19	20
21	22	23	24	25	26	27	28	29	30
31	32	33	34	35	36	37	38	39	40
41	42	43	44	45	46	47	48	49	50

	2	3	4	5	6	7	8	9	10
11	12	13	14	15	16	17	18	19	20
21	22	23	24	25	26	27	28	29	30
31	32	33	34	35	36	37	38	39	40
41	42	43	44	45	46	47	48	49	50

All of the numbers not eliminated are prime.

	2	3	4	5	6	7	8	9	10
11	12	13	14	15	16	17	18	19	20
21	22	23	24	25	26	27	28	29	30
31	32	33	34	35	36	37	38	39	40
41	42	43	44	45	46	47	48	49	50

Well done! You just found all the prime numbers between 1 and 50.

Connection

Mathematicians find that prime numbers are fun to think about in theory. However, it turns out that prime numbers also have a lot of modern, practical uses. Though any are worth studying, we will examine only one here.

Cryptology is the study and application of encrypting and decrypting, or encoding and decoding data. One of the earliest recorded *ciphers* is known as the Caesar Cipher. Used by Julius Caesar, this method encrypts a message by replacing one letter systematically with another. Using a (-3) cypher, each letter is replaced with the one that occurs 3 *before* it in the alphabet. Hence, "E" is replace be "B."

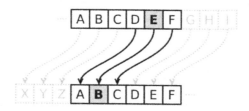

In this process, the message "THIS IS A SECRET" is encrypted as "QEFP FP X PBZOBQ." This cypher has two obvious weaknesses : (1) With enough messages of this type, we could do a frequency analysis to see which letters come up most often. These we could easily assign to T, S, R, and E. Then we could figure out the decryption. (2) Even withouta frequency analysis, if we knew how this was encrypted (-3), then we know how to decrypt the message (+3).

```
P │L–A│ Y  F         OL
I │ R E│ X  M     Shape: Rectangle
B │ C D│ G  H     Rule: Pick Same Rows,
K │N–O │ Q  S        Opposite Corners
T   U   V  W  Z      NA
```

Throughout time, people have developed many methods to encrypt messages. For instance, the Playfair Cipher was popularized by Disney's *National Treasure* movies. The intrigue and decryption of the Zimmerman Telegram and "AF is short of water" played important roles in the history of the twentieth century. World War II movies have romanticized the German *Enigma* machine, which mechanically encrypted messages. Its encryption was mistakenly said to be unbreakable.

Each of these ciphers, and many others, had the same weakness: if you knew how it was encrypted, it was easy to decrypt. The challenge for late twentieth century mathematicians was to find an encryption method that would allow the encryption key to be *public* while the decryption key remained *private*.

RSA Encryption is the most popular *public key encryption* system in use today. **R**ivest, **S**hamir, and **A**dleman of MIT developed this encryption system in 1977. You may have seen the RSA symbol on the bottom of webpages involving the sending of private information like credit card numbers. RSA encryption allows the public to know and use the encryption algorithm. However, unlike most other encryption approaches, just because we know how to encrypt the message does not mean we can figure out how to decrypt the message. This fact is what makes RSA so useful and powerful. Can you guess the basis for RSA encryption? Prime numbers!

More on Prime Numbers

You might think that since prime numbers have been studied by mathematicians for millennia, most questions involving primes would have been resolved by now. However, part of the allure of primes is that although they are easy to discuss, their behavior is in many ways quite literally unpredictable. To illustrate this point, let's consider four prime number topics.

The number of primes. Is there a finite or an infinite number of prime numbers? The proof that there are an infinite number of prime numbers goes back to Euclid's *The Elements* (300 BC). The list of prime numbers goes on forever. This means we can never have full list of primes.

The distribution of primes. So, we cannot make a list of prime numbers because they are infinite. We could say that of the even numbers, too. The difference is that we know how to predict exactly where we would find the even numbers. In other words, we understand the distribution of even numbers—every other number will be even. The same cannot be said of prime numbers.

The following series of questions illustrates this point rather simply. How many positive even numbers are there less than or equal to 10? Less than or equal to 100? 1000? 10,000? The answers are very easy to figure out; we could even develop a formula to predict the answer every time: $\frac{1}{2}(N)$.

N	Number of positive *even numbers* less than or equal to N
10	5
100	50
1000	500
10,000	5,000
K	0.5 K

What if we asked similar questions about the distribution of prime numbers? When and where do they occur? With some work, we can generate the related table. The function for the number of primes less than a number n is given the name $\pi(n)$ (read that "pi of n" not "pi times n"). There are 4 primes less than 10 and there are 168 primes less than the number 1000. , Consequently, $\pi(10) = 4$ and $\pi(1000) = 168$. Is there a formula for $\pi(n)$? At age 15, Gauss conjectured that $\pi(n) \approx \frac{n}{\ln(n)}$ where $\ln(n)$ is the natural logarithm of n. He later refined and proved his conjecture. Although his formula works as an approximation, mathematicians today are still pursuing a more exact formula.

N	Number of primes less than n $\pi(n)$
10	4
100	25
1000	168
10,000	1229

Twin Prime Conjecture. Within the 10 years before the writing of this book, mathematicians have made significant, notable progress related to the Twin Prime Conjecture. This is a very old conjecture that is easy to understand but very difficult to prove. Notice that it is a conjecture and not a theorem because it has not been proven.

You may have noticed that, except for 2 and 3, the closest two primes occur is two numbers apart. When this happens, we say those two primes are "twins." Pairs of twins include 5 and 7, 17 and 19, 41 and 43, 101 and 103, and 1997 and 1999. Mathematicians have noticed that no matter how far we calculate prime numbers, every once in a while a pair of twins appears. The Twin Prime Conjecture is that twin primes will continue to appear forever; it states that there are an infinite number of twin prime pairs. Simple, right? Yet mathematicians continue to puzzle over the proof of this conjecture. Why? Primes are easy to explain but difficult to bridle.

The largest known prime. Large prime numbers are both academically and practically significant. Practically, RSA encryption uses large prime numbers (200-400 digits long) to encrypt data. Academically, the pursuit of large primes has engaged both the professional and the amateur mathematician. A certain class of primes known as Mersenne Primes take the form $2^n - 1$. For example, $2^3 - 1 = 7$ is a Mersenne Prime. Not all Mersenne numbers are prime, but they are a good place to check for large prime numbers. Through the Great Internet Mersenne Prime Search (GIMPS), people have devoted the computing power of their computers to searching for large primes. At the writing of this book, the largest know prime is the 48th Mersenne Prime found. It is the number $2^{57,885,161} - 1$. The number has 17,425,170 digits. This is not a number just above 17 million—it has over *17 million digits* or, in scientific notation, is larger than $1 \times 10^{17,000,000}$. This is a really, really large number as compared to our experiences; the famous number googol is a 1 followed by only 100 zeros—only 101 total digits.

Something to Think About

Consider these three ideas: (1) prime numbers are the basic building blocks of arithmetic, (2) there are infinite primes and the largest know prime is over 17 million digits long, and (3) all current estimates of the number of elementary particles in the universe are under 100 digits long. These ideas make me wonder, and in this case I am using "wonder" in the "awe" sense of the word, at prime numbers. These basic building blocks of arithmetic, which is the foundation for mathematics, go so far beyond the natural world that I wrestle with the relationship between the natural, created universe and the study that is called mathematics. At some level, there is no way that the basic building blocks of mathematics describe the natural world. So what did God create, and what did we create?

Covering the Reading

1. Illustrate the Fundamental Theorem of Arithmetic for the numbers 100, 120, and 166.

2. Construct and apply the Sieve of Eratosthenes to find the prime numbers less than 200.

3. Use this chapter and a calculator to complete the following table:

N	$\pi(n)$	Gauss' first approximation for $\pi(n) = \dfrac{n}{\ln(n)}$	Difference between $\pi(n)$ and Gauss' approximation $\pi(n) - approx.$
10	4	4.34	-.34
100		21.71	
1,000	168		
10,000			

4. Explain what twin primes are, and give an example *not* listed in this chapter.

5. Explain how prime numbers could be considered the basic building blocks of mathematics.

Problems

6. Locate and read a few versions of the proof that there are an infinite number of primes. Write a summary of the proof.

7. In this chapter the application of primes to RSA encryption was presented. Find 2 other modern day, practical applications involving prime numbers.

8. Explore Goldbach's Conjecture. Explain what it is, and illustrate it.

9. Find 5 other theorems or conjectures about prime numbers. Explain them. If you find some you cannot explain, skip them until you get 5. Be sure to identify your findings as theorems or conjectures.

10. As part of the problems in this text, you have written about the relationship between God, humans, and mathematics. Reconsider and rewrite what you have said in the past to specifically focus on prime numbers. That is, explain the relationships between God, humans, and primes numbers.

11. Some prime numbers, the basic building blocks of arithmetic, are so large that they go beyond describing anything in the universe. What is the relationship between nature and mathematics? What did God create and what did we create, if anything?

Chapter 3.3
The Real Number Line, π, and e

This chapter examines the real number line with a specific focus on the numbers π and e. In order to understand the real numbers better, we will also study different portions of the real number line. As a brief review, the natural numbers, or counting numbers, are those you naturally count: $\{1, 2, 3, 4, 5, ...\}$. The integers are the natural numbers along with 0 and the opposites of each counting number: $\{...-3, -2, -1, 0, 1, 2, 3, ...\}$. The real numbers are any number that you can plot on the real number line:

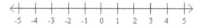

Warm-up Activity
Which numbers can you write as a fraction of two integers?

Decimal	Fraction	Decimal	Fraction
0.5	$1/2$	0.5472	
17		$0.\overline{47}$	
1.92		$0.3\overline{6}$	
$0.\overline{3}$		$\sqrt{2}$	

Concept Development

In the previous activity, all but one of the numbers can be written as a fraction. Any real number that can be written as a fraction of two integers is called a *rational* number. Any real number that cannot be written as a fraction of two integers is called an *irrational* number. There are a few tricks one can use to convert decimals to fractions. However, whether in decimal form, 0.5, or in fraction form, $\frac{1}{2}$, the number is still a rational number because it *can* be written as a fraction of two integers. For instance, the terminating decimal 1.92 can be written as $\frac{192}{100}$. Most people have memorized that the repeating decimal $0.\overline{3} = \frac{1}{3}$. It is a small manipulation to make this happen. A similar method works for all repeating decimals.

Let
$$x = 0.333333333\ldots$$
Then
$$10x = 3.333333333\ldots$$
Subtracting the two equations
$$10x = 3.3333333333\ldots$$
$$-(x = 0.333333333333\ldots)$$
Results in
$$9x = 3$$
Dividing both sides by 9 and reducing the fraction
$$x = \frac{3}{9} = \frac{1}{3}$$
So
$$x = 0.333333333\ldots \text{ and } x = \frac{1}{3}$$
Therefore
$$0.\overline{3} = \frac{1}{3}$$

Any terminating decimals (such as 0.42736) and any repeating decimals (such as $0.\overline{2561}$) can be written as a fraction; they are rational numbers. Consequently, irrational numbers are those that are not terminating and are not repeating. In decimal notations, irrational numbers appear as non-terminating, non-repeating decimals. Here are three examples:

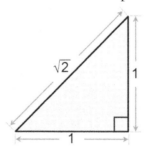

$\sqrt{2}$ is irrational.
$\sqrt{2} \approx$ 1.414213562373095048801688724209698078 ...
The decimal notation never ends and never falls into a repeating pattern.

π is irrational.
π ≈ 3.1415926535897932384626433832795028841 ....
The decimal notation never ends and never falls into a repeating pattern.

0.101001000100001000001000001 ... is irrational.
The decimal notation never ends and never falls into a repeating pattern. This one is a little different because its number pattern is predictable, but it is still irrational.

Thought Experiment
Suppose you used a 10-sided die with the ten digits 0, 1, 2, 3, 4, 5, 6, 7, 8, and 9 on the ten sides. You sit down with a lot of time and a large sheet of paper. You write, "0." on the paper

and then proceed to roll the die to determine digits after the decimal point. On your first roll you roll a 5. You write down the digit, so the paper now says, "0.5." The next three rolls are 3, 3, and 8 respectively. Your paper now says, "0.5338." If at any point you stop this process, you will have a terminating decimal and consequently, you will have recorded a rational number.

However, this is a thought experiment, so we can take certain liberties with reality. Suppose you continued rolling and recording forever. There are two possibilities: you will either record a repeating decimal (some sequence of numbers appears over-and-over again forever), or you will record a non-repeating decimal. By the way, you could get a terminating decimal if from some point on you roll the digit 0 forever. This thought experiment accounts for every possible decimal: terminating and repeating in one category and non-repeating in the other—rational numbers and irrational numbers, respectively.

What is the chance you will roll a repeating decimal? What is the chance your number will fall into some pattern that repeats over and over again forever? It is fairly intuitive to judge that although this *could* happen, it probably will not happen. You *could* randomly roll a rational number between 0 and 1, but you probably will not. Most likely, you will roll a non-repeating, irrational number between 0 and 1.

The die simulation is a way of *randomly* generating a number between 0 and 1. If we randomly generate a number between 0 and 1, how likely is it that we will get a rational number? 1 out of 10 times? 1 out of 100 times? Most people agree that it is more like 1 out of 1,000,000,000 times! If you don't, then you may want to reread the experiment because you may not understand it. It is highly unlikely, almost impossible, that you will roll a repeating pattern forever without it ever varying.

What does this thought experiment tell us? If we randomly select a number on the number line, the chance of it being rational is practically 0%, and the chance of it being irrational is practically 100% (or closer than we can possibly imagine). Most numbers on the real number line are irrational. Most numbers on the real number line *cannot* be written as a fraction of two integers.

This realization can be a bit unnerving; you have spent most of your life studying whole numbers and fractions: rational numbers. Yet most real numbers are not rational numbers. To put it bluntly, with respect to real numbers, you have learned quite a bit about practically nothing while learning practically nothing about most real numbers. By the way, you are in good company. The situation is not much better for mathematicians!

Going Deeper

You now realize that the real number line consists of mostly irrational numbers. These irrational numbers can be further subdivided into two categories: algebraic irrational and transcendental irrational numbers. Algebraic irrational numbers are irrational numbers that can be solutions to polynomial equations with rational coefficients. For example, consider this polynomial:

$$x^2 - 2 = 0$$

The two solutions to this polynomial are $\sqrt{2}$ and $-\sqrt{2}$. Therefore, $\sqrt{2}$ and $-\sqrt{2}$ are algebraic numbers. Furthermore,

$\sqrt{2}$ *and* $-\sqrt{2}$ cannot be written as a fraction of two integers. They are algebraic irrational numbers.

Mathematicians have demonstrated that the numbers π and *e cannot* be solutions to any polynomial equation with rational coefficients. The numbers π and *e* are transcendental irrationals; they transcend algebra. Most of the study of transcendental irrationals exceeds the scope of this text. However, it is important to note that among the irrationals, the algebraic ones and transcendental ones are not evenly divided. In fact, the same way most of the real numbers are irrationals, most irrational numbers are transcendental.

Basic arithmetic based on the integers and most of high school algebra lead us to a better understanding of rational and algebraic numbers, but most of the real numbers transcend algebra. The world of mathematics, even the basic world of the real number line, is vast. Humanity has only begun to reign in the mathematics of number.

An Introduction to the Number π

While we study the real number line, a few detours beckon us: the two specific transcendental irrational numbers π and *e* While almost everybody has heard of π, *e* is not as popular.

When I ask most people, "What is π?" the typical answer is "3.14." Although this is not entirely incorrect, it most definitely is not entirely correct. Most people know that 3.14 is just the start of a long line of digits used to write π. Understanding that π is an irrational number, you should expect that long expansion as well. It is more appropriate to say π is approximately 3.14. However, most people do *not really* answer the question, "What is π?" Many people think mathematicians decided to make up a number π and make it equal about 3.14. In reality, the story actually goes in the opposite direction.

Hypothetically, suppose you were to spend a month or a year or a lifetime walking around with a tape measure, measuring every single circle you see as accurately as you can. You measure the *circumference* of each circle (the distance around the circle) and you measure the *diameter* of the circle (the distance across the circle). While this may become incredibly tedious, you start to notice a pattern. The circumference of the circle is *about* three times as large as the diameter of the same circle. Your curiosity compels you to investigate this relationship more closely. What is the *factor* between the circumference and the diameter? You decide to take every circle measurement you have ever found and begin dividing: circumference divided by diameter. Every time you repeat this procedure, you get approximately the same number: 3.1415. No matter how big or how small the circle is, the ratio of the circumference to the diameter is about the same number.

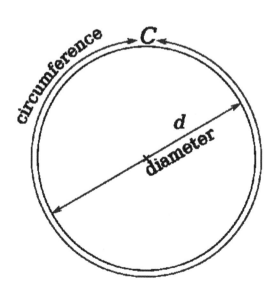

Hypothetically, you begin to investigate and it turns out that every person in every culture in every location around the world who performs these measurements finds the same results. Circumference divided by diameter equals about 3.1415. You and all of the other circle measurers decide to hold a conference. Everyone recognizes that this relationship is universal. That number that always describes the circumference divided by diameter needs a special name. You decide to hold a "Name that Number" contest. People from all over nominate names, and everyone votes. π wins. The ratio of the circumference to the diameter in any circle is π.

$$\frac{Circumference}{Diameter} = \pi$$

The fact that the Greek letter "pi" is a homophone to the English word "pie" naturally leads to a number of jokes and comics, as well as the making and consuming of many pies on March 14 (3-14) every year: pi day.

Approximating π. Many mathematicians across many cultures have pursued a better and better approximation for π. One of the earliest excellent approximations was developed by Archimedes (3rd century B.C.). His method was simple and insightful. It was based on the idea that calculating the perimeter of a regular polygon is fairly straightforward. Archimedes then used the perimeter of regular polygons to approximate the circumference of a circle.

How did he do this? Archimedes inscribed a circle inside a hexagon and a smaller hexagon inside the circle. The circumference of the circle would be between the perimeter of the larger and smaller hexagons. The hexagons also allowed him to easily calculate the diameter of the circle. Using these numbers, Archimedes found an approximation for π.

However, still unsatisfied, Archimedes then found an even better approximation for π by doubling the number of sides of the polygon. Using a dodecagon, he improved the approximation. Even though the dodecagon already almost looks like a circle, Archimedes continued to using 24-gons, 48-gons, and finally 96-gons to get an excellent approximation for π. His approximation was so good that today, over two thousand years later, it is still sufficient for almost all modern physics applications.

Over time, many people and cultures have worked on improving the approximation for π; today, many methods can be used to find a better and better approximation for π. One method can be applied using a modern calculator and an infinite series developed as part of calculus:

$$\pi = 4\left(1 - \frac{1}{3} + \frac{1}{5} - \frac{1}{7} + \frac{1}{9} - \frac{1}{11} + \cdots\right)$$

An Introduction to the Number e

The number e is also a transcendental irrational number. The number $e \approx 2.718$, and is known as Euler's Number (not to be confused with Euler's Constant). Euler (pronounced "oiler") was a mathematically prolific 18th century Swiss mathematician who was a professing follower of Jesus Christ. Euler's number is crucial to many useful equations across many disciplines. Here are some examples:

Continuous compounding interest (economics) and exponential population growth formula (sociology): $A = Pe^{rt}$

Logistic population growth models (when there are limited resources, usually in biology):
$$A = \frac{KPe^{rt}}{K + P(e^{rt} - 1)}$$

Newton's Laws of Heating and Cooling (forensics- how long has this murder victim been dead?):
$$Temperature\ at\ time\ t = Ambient\ Temp + (Initial\ Temp - Ambient\ Temp)e^{-kt}$$

Free fall velocity with air resistance (physics):
$$v = \sqrt{\frac{mg}{k}} \times \left(\frac{1 - e^{-2\left(\frac{t}{\sqrt{\frac{m}{gk}}}\right)}}{1 + e^{-2\left(\frac{t}{\sqrt{\frac{m}{gk}}}\right)}} \right)$$

Calculation for piano frequencies for Bach's Well-Tempered Clavier (music):

$$r = ae^{b\theta}$$

Music and the Number *e*. Piano tuners do not really use that last example; they do not sit at a piano and make calculations with the above formula. However, the application is intriguing and relates to another natural phenomenon. In the 1600 and 1700s, while the Bach family was dominating the world of music, the Bernoulli family was dominating the world of mathematics. Both families contributed significantly to their fields over multiple generations. However, the Bernoullis personal relationships were frequently abrasive. Unfortunately, the Bachs and Bournoullis did not get to communicate with each other. Why is this so unfortunate? The Bernoullis had a simple solution to a problem the Bachs and all musicians faced.

This problem was changing keys in the middle of a song. The 12-tone note system of music is based on ratios developed by the ancient Pythagoreans. For example, if you have two identical strings under the same tension and one is twice as long as the other, they will play notes exactly one octave apart. In other words, a 2:1 ratio results in an octave. All other notes are also in whole number ratios to each other. Consequently, the differences between individual notes are not *exactly* equal, but the notes sound absolutely perfect.

When a piece of music changes key, we have to maintain the perfection without re-tuning the instruments to the new key. The problem is that differences between consecutive notes played on the same instrument are off in the new key, and the music sounds bad. Focusing on the piano, Johann Sebastian Bach found a solution. He made all of the differences between consecutive notes equal. While the resulting music sounds just a tiny bit bad in every key, the musician can change keys as often as desired. Bach's solution was still an acceptable compromise.

Where do the Bernoullis come in to the story? Actually, they never do. However, they could have offered the logarithmic spiral: $r = ae^{b\theta}$. The logarithmic spiral would have calculated exactly what Bach needed with great ease. With the logarithmic spiral, the precision Bach needed could be calculated as easily as adding $1 + 1$.

The other bizarre link between music, Euler's number, and natural phenomena is that the logarithmic spiral also most easily models the Golden Spiral described in the chapter on phi: the Golden Ratio. This golden ratio is supposedly found many natural phenomena.

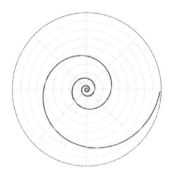

What is the number *e*? Although we have been examining the applications of the number *e*, we have not yet analytically defined it. We will end the concept development of this chapter with two different expressions for Euler's number.

Due to its transcendental, irrational nature, e is most easily defined with infinity. It can be written either as a limit or as an infinite series:

$$e = \lim_{n \to \infty} \left(1 + \frac{1}{n}\right)^n$$

$$e = 1 + \frac{1}{1!} + \frac{1}{2!} + \frac{1}{3!} + \frac{1}{4!} \cdots$$

(note: 4! is not an excited 4. 4! reads "four factorial" and means 4x3x2x1.)

Something to Think About

In this chapter we have taken a short tour of the real number line. We explored rational and irrational numbers, algebraic and transcendental numbers, and the specific numbers π and e. This growth in understanding expands our specific understanding of mathematics. However, we also need to consider a few other thoughts related this chapter.

First, recall that practically all of the real numbers are irrational: they cannot be expressed as fractions of two integers. Conversely, almost none of the real numbers are rational numbers. Yet almost all of the numbers we express in our daily experience as humans are rational. If there are so many more irrationals than rationals, why do we as humans tend to only use rationals?

Secondly, the number π is approximated in the Bible. As part of the construction of the temple in Jerusalem, Solomon made a round bowl described in 2 Chronicles 4:2. "Also he made a molten sea of ten cubits from brim to brim, round in compass, and five cubits the height thereof; and a line of thirty cubits did compass it round about" (KJV). In other words, the round bowl in front of the temple was 30 cubits around and 10 cubits across. Using these numbers $\pi = \frac{30\ Cubits}{10\ Cubits} = 3$, critics claim that this statement that $\pi = 3$ in the Bible proves that the scriptures contain errors. After all, here is a clear error. Does this verse threaten the claim that the Bible is inerrant and make its inspiration suspect?

Solomon Dedicates the Temple at Jerusalem
James Jacques Joseph Tissot

Finally, Euler's number has many applications when *natural* phenomena are involved and consequently might hold a unique place among numbers. It might be possible to say that e is truly a special number. Are some numbers more special than others? What would make e more special than other numbers?

Mathematical Explorations

Covering the Reading

1. Write 0.515 and $0.\overline{45}$ as fractions.

2. Write a summary contrasting the rational and irrational numbers. Include at least 3 contrasts.

3. What makes a number transcendental?

4. What is π?

5. Approximate π using $4\left(1 - \frac{1}{3} + \frac{1}{5} - \frac{1}{7} + \frac{1}{9} - \frac{1}{11}\right)$.

6. Approximate e to 7 decimal places using one of the expressions in the chapter.

Problems

7. In your own words, argue that there are more irrational numbers than rational numbers on the real number line.

8. Find and measure 5 circles. Use the measurements to estimate π.

9. Find another equation that involves the number e.

10. If there are so many more irrationals than rationals, why do we as humans tend to only use rationals? Explain how you come to your conclusion.

11. Respond to the claim that the Bible is in error when 2 Chron. 4:2 indicates that $\pi = 3$.

12. It might be possible to say that e is truly a special number. Are some numbers more special than others? What would make e more special than other numbers?

Chapter 3.4
Counting Infinity

Infinity is a concept, not a number. Since this is the case, an infinite set cannot be counted or measured in the same way a finite set, or finite number of things, can be counted. We can accept these statements because we realize that the finite does not behave like the infinite. Yet the challenge in this chapter will be *to avoid* applying our finite understanding to infinity. We will constantly want to grasp onto our finite experiences and apply them to the infinite. While some finite principles so apply when dealing with the infinite, we will have to abandon our experiences and allow pure reason to override "common sense." Let's go!

Warm-up Activity

1. Are there more positive whole numbers from 1 to 100 or more positive even numbers from 1 to 100? How do you know? Can you explain without actually counting them? Are there more positive whole numbers from 1 to 1,000 or more positive even numbers from 1 to 1,000? How do you know? Can you explain without actually counting them? Are there more positive whole numbers from 1 to 1,000,000 or more positive

even numbers from 1 to 1,000,000? How do you know? Can you explain without actually counting them?

2. Suppose you are organizing an open-house event in a large room arranged with rows of chairs. Ten minutes after opening the doors to the room, some people are sitting while many are milling around talking with each other. You suddenly realize that there may not be enough chairs for everyone. You have no idea how many chair or people there are. *Without counting*, how can you determine if there are more chairs or more people?

3. What does it mean for a set of numbers to be infinite? Try to define the term *infinite set*.

Concept Development

Learning from Warm-up Part 1. In Warm-Up number 1, you probably came to the conclusion that there are more positive whole numbers than positive even numbers over any interval. No matter how large the interval, this is always true. However, what happens if we change the question to involve infinity? Are there more positive whole numbers or positive even numbers? This question can truly vex us because the answer could be argued two different ways.

We could argue that there are twice as many positive whole numbers as positive even numbers for three reasons. 1. The Pattern Argument: It is true for 100. It is true for 1,000. It is true for 1,000,000. Consequently, the trend *must* continue to infinity, right? Actually, this is a case of potentially misapplying the finite to the infinite. 2. The Piles of Numbers Argument: Another argument we could make is that there are twice as many positive whole numbers as positive even numbers. For instance, if we were to put the numbers into two piles, then we would put two numbers into the "whole number pile" for each time we put one number into the "even number pile." This is basically the same argument as the first one. The question that remains is also the same: "does this finite experience applies to the infinite?" 3. The Leapfrog Argument: Finally, we could contend that the list of evens skips every other whole number. Therefore, the evens equal the whole numbers *minus* something: the odds. The potential trouble here is this: in every finite experience, if you take something away from the whole, then the result is something smaller. Is this property true of infinite sets? If you remove something from an infinite set, is the result always smaller?

If this were a conversation, this is the point at which some people just waive their hands and say, "Infinity is infinity. So they are both just infinite." However, we will learn in this chapter that this statement oversimplifies the problem. Infinity is not always the same. We will consider the construct of different infinities (yes, plural—more than one). In turn, this allows us to consider whether some infinities are bigger than others.

Learning from Warm-up Part 2. In Warm-Up 2, you consider the people-and-chairs problem. How can you determine if there are more chairs or people in a room? The simple, non-counting solution is to ask everybody to "Please, sit down." Mathematically, what you are trying to do is make a *one-to-one correspondence* between people and chairs. After everyone sits down, you can easily tell if there are more chairs, more people, or if the numbers are equal.

This concept of one-to-one correspondence allows us to determine relative sizes of sets *without* having to actually count them. This method allows us to see if Set A is larger than Set B, if Set B is larger than Set A, or if the two sets are the same size. This is the method employed by classical mathematical analysis to handle the questions of infinite sets. With finite sets, there are two ways to determine which set is bigger: counting and one-to-one correspondence. However, since infinite sets cannot be counted, the only method that remains is one-to-one correspondence.

One-to-one Correspondence

How does one-to-one correspondence work with infinite sets? We will examine this concept two ways. First, we will introduce working with infinity and one-to-one correspondence using "Hilbert's Hotel," And second, we will construct a number of one-to-one correspondences.

Hilbert's Hotel. David Hilbert was a German mathematician who lived from the mid-1800's to the mid-1900's. He famously posed 23 unsolved problems at the 1900 Paris conference of mathematicians. These 23 unsolved problems in mathematics guided many mathematical endeavors of the twentieth century. As you may suspect, Hilbert was also famous for his work in mathematical analysis including infinite sets.

The so-named Hilbert's Hotel is a synthetic construct: suppose there was an infinite hotel with an infinite number of rooms numbered 1, 2, 3, 4, ... One night, an infinite number of people show up at the hotel and fill up the rooms. We will say person P7 is in room R7. Should the hotel clerk turn on the "No Vacancy" sign? Consider what happens when someone else shows up at the hotel at this point.

Suppose person A shows up at the hotel. Is there room for person A? If we were in a finite hotel, the answer would be "no." However, this is Hilbert's Hotel. We could try to assign person A to the "last room." However, at Hilbert's Hotel, there is no "last room." What does the hotel clerk do? Fortunately, the clerk is mathematically literate. The clerk asks person P1 to move to room R2, person P2 to move to room R3, person P3 to move to room R4, and so on. In other words, person P(n) moves to room R(n+1). This leaves room R1 open for person A and everyone will still have a room.

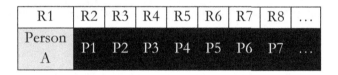

Now, you may object, "Where will the last person go? Won't they be pushed out of the hotel?" The answer would be "yes" if this were a finite hotel, but it is not. This hotel has an infinite number of rooms, so we can always add one more.

Using the same line of reasoning, we should manage to fit not only one more person but 10 more people or 1,000 more people into the hotel. Let us return to the original situation where person P1 is in room R1. If 1,000 more people showed up, the clerk would simply ask person P1 to move to room R1001, P2 to move to room R1002, P3 to move to room R1003, and so on. Person P(n) moves to room R(n+1000). Then there would be 1,000 open rooms for the 1,000 people who showed up. Any finite number of people who show up can be accommodated in Hibert's Hotel, even when it is full.

However, since we are playing these infinity games, suppose an infinite bus arrives at Hilbert's Hotel. The people on the infinite bus are H1, H2, H3, … If we return to our original scenario where P1 is in room R1, is there room for everyone on the infinite bus? The clerk tells the group on the bus, "We have room for you at Hilbert's Hotel." The clerk proceeds to assign person P1 to room R2, P2 to room R4, P3 to room R6, P4 to room R8, P5 to room R10, and so on. Person P(n) moves to room R(2n). Consequently, all of the even rooms are filled by persons P1, P2, P3, … while those from the bus, H1, H2, H3, …, are assigned to the odd rooms. Person H1 to R1, H2 to R3, H3 to R5, H4 to R7, and so on. Person H(n) is assigned to room R(2n-1). Every person has a room, and every room has a person.

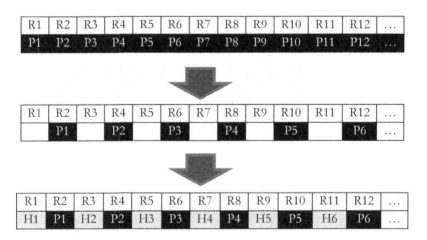

Now we have seen that an infinite bus can show up, and there is still room at Hibert's Hotel. The next challenge is this: Suppose an infinite number of infinite busses, B1, B2, B3, B4,…, arrive at Hibert's Hotel. Is there room at the hotel? In other words, can we fit Bus 1 Human 1 (B1H1), B1H2, B1H3, …, and

B2H1, B2H2, B2H3,..., and B3H1, B3H2, B3H3,..., and so on into the hotel?

Constructing one-to-one correspondences. Hilbert was not the first mathematician to work on these types of problems; the hotel is simply named after him. The difficult work of answering these questions was tackled by an eighteenth-century German mathematician named Georg Cantor, who worked in set-theory and developed a theory of trans-finite numbers. If the mathematics that we have been doing in this chapter does not sit well with you, then you are not alone. While developing the mathematics of infinite sets, Cantor's work was severely critiqued by accomplished mathematicians Henri Poincaré and Leopold Kronecker. Between wrestling with infinity and enduring harsh criticism, Cantor found himself in and out of a mental institution in Halle, Germany.

Cantor was the one who found a way to fit the infinite people on the infinite busses in the infinite hotel. However, when he looked at the problem, it wasn't explained through a hotel scenario; it had to do with sets of numbers. His scheme uses one-to-one correspondence with the set of what are called natural numbers or counting numbers. The set of counting numbers is $\{1, 2, 3, 4, 5, ...\}$. The mission was to determine which sets of numbers could be set in a one-to-one correspondence with the counting numbers. Any set that can be set in one-to-one correspondence is called *countable*. Countable does not mean that it can be counted, but that it matches with the counting numbers. Any sets that can be *mapped* to the counting numbers are the same size, or *cardinality*, as the counting numbers. Furthermore, if two sets can be mapped to the counting numbers, then they can be mapped to each other; they are the same cardinality. Here are some examples of countable sets.

The set of positive even numbers are in one-to-one correspondence with the counting numbers:

$$1 \leftrightarrow 2$$
$$2 \leftrightarrow 4$$
$$3 \leftrightarrow 6$$
$$4 \leftrightarrow 8$$
$$5 \leftrightarrow 10$$
$$6 \leftrightarrow 12$$
$$7 \leftrightarrow 14$$
$$8 \leftrightarrow 16$$
$$9 \leftrightarrow 18$$
$$\vdots$$

Consider this correspondence in terms of the people-and-chairs example at the start of the chapter. If the set of counting numbers, {1, 2, 3, 4, 5, ... }, are people and the set of positive even numbers, {2, 4, 6, 8, 10, ... }, are chairs, then every person has a chair and every chair has a person. There are an equal number of people and chairs; the set of counting numbers is the same cardinality as the set of positive even numbers.

Take a moment to return to Warm-Up 3. This correspondence also helps to illustrate the definition of an *infinite set*. How did you define infinite set? Most people use phrases like "never-ending" or "go on forever." However, mathematicians have found these types of definitions to be insufficient. Consider this following one. In any *finite set*, a piece of the set is always smaller than the whole set. More rigorously, a proper subset of a finite set is smaller in size than the whole set. However, in the one-to-one correspondence above, while the set of positive even numbers is a subset of the set of counting numbers, they are the same cardinality. In other words, a part is just as big as the whole. You have now reached the mathematical definition of an infinite set. A set is infinite if a proper subset can be mapped in a one-to-one correspondence to itself. The part is as big as the whole. This definition again illustrates that our experience with the finite, in which the part is smaller than the whole, does not always apply to the infinite, where the part can be as big as the whole.

Let's continue considering countable sets. The set of integers {..., −3, −2, −1, 0, 1, 2, 3, ... } is also in one-to-one correspondence with the counting numbers. However, the mapping is a bit more challenging since, like a line instead of a ray, the set of integers continues infinitely toward both positive and negative infinity. However, the sets can be set in one-to-one correspondence; the set of integers is countable. The key is that we need to hop back and forth from positive to negative numbers. Basically, the mapping is the same as the solution to Hilbert's Hotel and the arrival of the infinite bus. Zero and the positive numbers are mapped to the odds while the negatives are mapped to the evens.

$$
\begin{aligned}
1 &\leftrightarrow 0 \\
2 &\leftrightarrow -1 \\
3 &\leftrightarrow 1 \\
4 &\leftrightarrow -2 \\
5 &\leftrightarrow 2 \\
6 &\leftrightarrow -3 \\
7 &\leftrightarrow 3 \\
8 &\leftrightarrow -4 \\
9 &\leftrightarrow 4 \\
&\vdots
\end{aligned}
$$

Every person has a seat, and every seat has a person.

Two Major Challenges

Through these one-to-one correspondences, we have seen that the cardinalities of the positive even numbers, the integers, and the counting numbers are all the same. All the sets are said to be *countable* because they can be mapped in a one-to-one correspondence with the counting numbers. It is not difficult to map the whole numbers, the multiples of five, the multiples of seven, or even the Fibonacci numbers to the counting numbers. All of these sets are countable. However, particular challenges arise when we consider the cardinality of the rational numbers and the real numbers.

Rational Numbers. Remember, rational numbers are any numbers that can be written as a fraction of two integers, they include fractions like $\frac{2}{3}, \frac{-5}{7}, 4 = \frac{4}{1}, \frac{202387}{33837}, and\ 0 = \frac{0}{8}$. Irrational numbers are numbers that cannot be written as a fraction, like $\pi, e,$ and $\sqrt{2}$. Now, "Can the rational numbers be mapped in a one-to-one correspondence with the counting numbers? In other words, are the rationals countable? Or are there categorically more rational numbers than counting numbers? Not surprisingly, Cantor devised a means to answer this question. For the sake of simplicity, we will look at the one-to-one correspondence between the positive rational numbers and the counting numbers (from this step, you can then map all of the rational numbers using the positives-to-evens, negatives-to-odds argument). To help understand the magnitude of the challenge, consider it in the context of Hilbert's Hotel, at the bus stage of the story. Basically, the challenge is to find hotel rooms for the arrival of an infinite number of infinite busses. Why? Consider bus number 5. Let's say Bus 5 represents all of the fractions with a denominator of 5. The bus is infinite because there are infinite numerators with the denominator 5. Then Bus 6 contains all the numbers whose denominator is 6, and so on. The number of busses is infinite since there are an infinite number of possible denominators.

Cantor's method for solving this problem has been named *Cantor's Diagonalization Process*. Its name provides a hint as to nature of the mapping. Let's work through it together. First, Cantor found a way to list the positive rationals to set himself up for success:

$$\begin{array}{cccccc}
1/1 & 2/1 & 3/1 & 4/1 & 5/1 & 6/1 & \cdots \\
1/2 & 2/2 & 3/2 & 4/2 & 5/2 & 6/2 & \cdots \\
1/3 & 2/3 & 3/3 & 4/3 & 5/3 & 6/3 & \cdots \\
1/4 & 2/4 & 3/4 & 4/4 & 5/4 & 6/4 & \cdots \\
1/5 & 2/5 & 3/5 & 4/5 & 5/5 & 6/5 & \cdots \\
1/6 & 2/6 & 3/6 & 4/6 & 5/6 & 6/6 & \cdots \\
\vdots & \vdots & \vdots & \vdots & \vdots & \vdots & \ddots
\end{array}$$

This set-up allows Cantor to list *every* positive rational number. You probably noticed that equivalent fractions like $2/4$ and $3/6$ are both listed. Since the challenge is to make sure that there are *not* more rational numbers than counting numbers, we have liberally listed every possible configuration of rational number to make sure we account for them all.

Prepared for success, Cantor proceeded to make a path through the rational numbers that was sure to pass through *every* number. As you can see below, this path clearly involves diagonals.

$$\begin{array}{cccccc}
1/1 & 2/1 & 3/1 & 4/1 & 5/1 & 6/1 & \cdots \\
1/2 & 2/2 & 3/2 & 4/2 & 5/2 & 6/2 & \cdots \\
1/3 & 2/3 & 3/3 & 4/3 & 5/3 & 6/3 & \cdots \\
1/4 & 2/4 & 3/4 & 4/4 & 5/4 & 6/4 & \cdots \\
1/5 & 2/5 & 3/5 & 4/5 & 5/5 & 6/5 & \cdots \\
1/6 & 2/6 & 3/6 & 4/6 & 5/6 & 6/6 & \cdots \\
\vdots & \vdots & \vdots & \vdots & \vdots & \vdots & \ddots
\end{array}$$

Notice that this path will cross over every rational number at some point—none are left out. This plan allows us to set up a one-to-one correspondence with the counting numbers. By following the path, we can map $1/1$ to 1, $2/1$ to 2, $1/2$ to 3, $1/3$ to 4, and so on:

$$1 \leftrightarrow 1/1$$
$$2 \leftrightarrow 2/1$$
$$3 \leftrightarrow 1/2$$
$$4 \leftrightarrow 1/3$$
$$5 \leftrightarrow 2/2$$
$$6 \leftrightarrow 3/1$$
$$7 \leftrightarrow 4/1$$
$$8 \leftrightarrow 3/2$$
$$9 \leftrightarrow 2/3$$

What do we conclude? Can we map the rational numbers to the counting numbers? Yes! Every person has a seat, and every seat has a person. The set of rational numbers is countable.

Real Numbers. The set of real numbers is every value on the number line. It therefore includes the rational numbers and the irrational numbers. In terms of decimal notation, real numbers include every number that could possibly be expressed. Remember, the rational numbers have a terminating or repeating decimal. For example, $7\frac{3}{4} = 7.75$ and $\frac{2}{3} = 0.\overline{6}$. On the other hand, irrational numbers do not have a terminating or repeating decimal: the decimal expansion for π goes on forever without falling into a repeating pattern. The real number set includes every possible decimal expansion.

Cantor's next challenge was to determine if he could find a one-to-one correspondence between the real numbers and the counting numbers. Are the real numbers countable or are they a completely different size of infinity?

Cantor ended up proving two things: he could not devise a one-to-one correspondence between the real numbers and the counting numbers, and that doing so is actually impossible. He used an indirect proof to show that the real numbers are not countable. I did not include that here, but you could easily find it online if you are curious. The real numbers are a completely different level or kind of infinity compared to all of the sets we have considered so far. While the counting numbers, positive even numbers, integers, and even the rational numbers are all the same size, as we saw by their one-to-one correspondence, the real numbers are a different size infinity.

Something to Think About

The concept of multiple infinities is usually new and uncomfortable for most people. If you feel that way, you are not alone. Considering the arguments Cantor made challenges our minds to think according to different rules. When we wrestle with the infinite, we cannot rely on our finite experiences as a reliable frame of reference.

The arguments involving the nature of infinite sets as they relate to reality can be traced back to Plato and Aristotle and even further back. Plato and Aristotle drew different conclusions about the nature of an infinite set. To Plato, infinite sets actually existed in our understanding. Aristotle was comfortable with the concept of *potential infinity*, but he asserted that since ultimate reality is grounded in our senses and actual infinity cannot be experienced, there is no such thing as infinity. The debate continues to this day. One thing is certain: whenever we wrestle with the nature of infinity, "religion" enters the picture. Some say that infinity exists in the mind of God, but not in the mind of man. What is infinity? Where is it? What is an infinite set? Do infinite sets exist? These are all difficult foundational mathematics issues.

After Cantor, a group of mathematicians later labeled as *intuitionists* rejected his work. They found Cantor's work with infinity so disturbing that they demanded that mathematics come back to more certain foundations. They did so by only accepting proofs that did not invoke infinity. For them to accept a piece of mathematics, it had to be arrived at through more concrete, finite steps. Their stance significantly reduced what could be considered valid mathematics, but it did provide them with what they considered a safe starting point. Infinite sets are definitely not safe mathematics.

Covering the Reading

1. Explain why the sets {a,b,c,d,e} and {1,2,3,4} *cannot* be mapped in a one-to-one correspondence.

2. What would the clerk at Hilbert's Hotel do if 15 people showed up and all the rooms were full?

3. Use the mappings described in this chapter to figure out which number would be mapped to the counting number 10 for the set of:
    a. Positive Even Numbers
    b. Integers
    c. Rational Numbers

4. Explain what it means for a set to be *countable*.

Problems

5. Show that the set of Whole Numbers $\{0,1,2,3,4,...\}$ is countable by defining a one-to-one correspondence.

6. Show that the set of multiples of 3 $\{...,-9,-6,-3,0,3,6,9,...\}$ is countable by defining a one-to-one correspondence.

7. This chapter used Cantor's Diagonalization Process to map the positive rational numbers to the counting numbers. Develop a similar process that will map ALL rational numbers to the counting numbers.

8. Find a proof that "The Real Numbers are *not Countable*" and explain the general idea in your own words.

9. Explain how your intuitions have been challenged in this chapter.

10. What impact does the idea that there are different sized infinities have on you? How does our study of infinity (so far) alter your understanding of our infinite God?

Chapter 3.5
Modular Arithmetic and Finite Groups

Whether you realize it or not, much of our study of arithmetic and algebra involves infinite sets. The sets of natural numbers and real numbers extend into infinity. Between any two real numbers, there are an infinite number of numbers. These concepts significantly challenge our intuitions; even those of us with the greatest computing abilities often struggle with thinking about infinity. This chapter proceeds in a completely different direction. When we add or multiply with the real numbers, we are working with a strong mathematical structure that is infinite. In this chapter, we will consider structures with few elements, yet the mathematical structures themselves will be as solid as the arithmetic and algebra you have used throughout your mathematical education.

Warm-up Activity

True or false
A. $1 + 1$ *is* 2
B. $2 + 2$ *is* 4
C. $4 + 4$ *is* 8
D. $8 + 8$ *is* 4

Concept Development

You should be suspicious of that Warm-up activity. After reading on, you will agree that all of the items are true. In fact, although you might think D is false, you have concluded that 8 + 8 *is* 4 many times in the past. You will also agree that 10 + 4 *is* 2. You do this all the time!

Let me re-emphasize that word: *time*. Instead of thinking about these items in the context of standard arithmetic, think about them on a clock. If it is 8 o'clock, then 8 hours later is 4 o'clock. 8 + 8 *is* 4. Similarly, if it is 10 o'clock, then 4 hours later is 2 o'clock. 10 + 4 *is* 2.

You may argue that what we are doing is not really arithmetic but you would be wrong. This is a type of arithmetic called *clock arithmetic*, which is itself a type of *modular arithmetic*. Modular arithmetic is not limited to just 12 elements; you can have any finite number of elements. For example, we can do arithmetic modulo 5. In modulo 5, the "clock" has only 5 numbers: 0, 1, 2, 3, and 4. (Notice the 0 instead of the 5. In standard arithmetic modulo 12, the 12 is replaced by 0 so that modulo 12 uses the numbers 0, 1, 2, 3, 4, 5, 6, 7, 8, 9, 10, and 11. Think of a clock with a 0 instead of a 12.)

Let's look at some examples of working in modulo 5. Notice that we replace the "equal to" sign, =, with a "congruent to" sign, ≡. I'll explain why later. In modulo 5, we repeatedly cycle through the numbers 0 through 4. Look:

$$0 \equiv 0 \ (mod \ 5)$$
$$1 \equiv 1 \ (mod \ 5)$$
$$2 \equiv 2 \ (mod \ 5)$$
$$3 \equiv 3 \ (mod \ 5)$$
$$4 \equiv 4 \ (mod \ 5)$$
$$5 \equiv 0 \ (mod \ 5)$$
$$6 \equiv 1 \ (mod \ 5)$$
$$7 \equiv 2 \ (mod \ 5)$$
$$8 \equiv 3 \ (mod \ 5)$$
$$9 \equiv 4 \ (mod \ 5)$$
$$10 \equiv 0 \ (mod \ 5)$$
$$11 \equiv 1 \ (mod \ 5)$$
$$12 \equiv 2 \ (mod \ 5)$$
$$13 \equiv 3 \ (mod \ 5)$$
$$14 \equiv 4 \ (mod \ 5)$$
$$3 + 3 = 6 \equiv 1 (mod \ 5)$$
$$3 \times 3 = 9 \equiv 4 \ (mod \ 5)$$
$$4 + 2 = 6 \equiv 1 \ (mod \ 5)$$
$$8 + 6 = 14 \equiv 4 \ (mod \ 5) \ or \ 8 + 6 \equiv 3 + 1 \equiv 4 \ (mod \ 5)$$

In the last example, you should notice that you can add then find the *congruence class* or you can find the congruence class then add. It makes no difference which method you use. This fact is part of the reason modular arithmetic is a strong mathematical structure.

The term *congruence class* should be a fairly accessible concept. Notice that in modulo 5, all of the following numbers are congruent to 3 $(mod \ 5)$:

$$\{\cdots, 3, 8, 13, 18, 23, 28, \cdots\}$$

All of these numbers are therefore in the same congruence class. They are all congruent to 3 and congruent to one another modulo 5. Any of them can be used interchangeably in modulo 5. The numbers 28, 8, 113, and even -2 are all congruent to 3 $(mod\ 5)$.

There are two ways to think of congruence classes. One way is that in any modulus n, any numbers that are n or a multiple of n apart from each other are in the same congruence class. Notice that all of the numbers in the set above are 5 or a multiple of 5 apart. This pattern includes negative numbers. Notice that -2 is 5 away from 3. Therefore, the congruence class containing 3 in modulo 5 is

$$\{\cdots -12, -7, -2, 3, 8, 13, 18, 23, 28, \cdots\}.$$

Using the idea of "n or a multiple of n apart," we can easily calculate larger numbers. For example, which congruence class contains the number 1517? As long as we add or subtract multiples of 5, we will find numbers in the same congruence class. $1517 \equiv 1517 - 1500 \equiv 17 \equiv 17 - 15 \equiv 2$. Why do we stop at 2? We typically represent the whole congruence class using the least non-negative number. 2 is called the *least positive residue*. We can represent modulo 5 with $\{0, 1, 2, 3, 4\}$. This set is called the *least residue system modulo 5*.

You may have figured out the second approach to finding the least positive residue. What happens when we divide 7, 12, 17, and 22 by 5? We have different quotients, but the remainder is always 2. Every element in the same congruence class has the same remainder. You can calculate the least positive residue either way.

You are probably wondering why modular systems are worth all this trouble. Do we use modular systems in daily life in other applications than when looking at a clock? Of course! As we discussed in the chapter on prime numbers, RSA encryption keeps our electronic data safe. RSA is based on prime numbers *within* modular systems. Check-digit systems also use modular arithmetic. The publishing industry uses modular systems for book recognition: ISBN-10 uses modulo 11 and ISBN-13 uses modulo 10 to ensure that the correct ISBN was scanned. Modular systems are also applied to UPC symbols and credit card numbers to verify correct data entry. In addition to these technological applications (and there are many more), we use modular systems in practical applications like the modulo 12 of the clock, modulo 7 of the days of the week, modulo 12 of the months, and modulo 4 of the seasons.

This introduction to modular arithmetic will allow us to consider finite mathematical systems. In modulo 5, for instance, all of the infinite number of integers can be reduced to $\{0, 1, 2, 3, 4\}$. This set along with an operation will have certain properties that make it mathematically strong.

Abelian Groups

In order to understand what an Abelian group is, you have to understand seven ideas. They are all embedded throughout the K-12 curriculum. You may have heard of them and identified them, but may not have really understood why you were learning them.

Set. The first idea is a *set*. We have been using this term throughout this book without defining it. At this point, you can probably provide a decent definition. We will casually call a set "any collection of elements." The alphabet is a set. $\{0, 1, 2, 3, 4\}$ is a set. The natural numbers are a set.

Binary Operation. A *binary operation* is any operation that takes two elements and provides a well-defined result. The four standard operations are $+, -, \times, and \div$. In a binary operation we put two elements in, like 4 and 2, and get an element in return.

$$4 + 2 = 6$$
$$4 - 2 = 2$$
$$4 \times 2 = 8$$
$$4 \div 2 = 2$$

There are also other, less common binary operations. For example, we could define $a \downarrow b$ to produce the lesser of the two values. Consequently, $5 \downarrow 4 = 4$ and $-2.5 \downarrow 2 = -2.5$. We could define $a * b = a^b$. In this case, $2 * 3 = 2^3 = 8$.

Commutative Property. With a given set and binary operation, there are 5 properties we will consider. The first property is the *commutative property*, and is familiar to most people. In words, the commutative property states that you can operate in either order and get the same result. Consider the set of natural numbers $\{1, 2, 3, 4, 5, \cdots\}$ with the operation $+$. $3 + 4$ is the same as $4 + 3$. The commutative property holds for all natural numbers with the operation $+$. However, considering the natural numbers and $-$, we see that $3 - 4$ is not the same as $4 - 3$. The commutative property can be symbolically represented as $a \blacksquare b = b \blacksquare a$ where $\blacksquare$ is any operation.

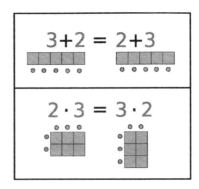

Closure Property. A set with a binary operation is considered *closed* if, when you operate two elements of the set the result you get is also in the set. Again, consider the natural numbers and normal addition. If we add any two natural numbers, the result is another natural number. However, if we subtract any two natural numbers, sometimes we get a natural number and sometimes we do not. $5 - 3 = 2$ but $7 - 11 = -4$, and -4 is not a natural number. Consequently, the natural numbers with addition are closed but the natural numbers with subtraction are not closed.

Identity Property. The *identity property* holds for a set with an operation if one of the elements produces an identical element. That element must do nothing. For example, consider the set of integers $\{\cdots, -3, -2, -1, 0, 1, 2, 3, \cdots\}$ with the operation of addition. Is there an element that we can add to any other element that produces the identical element? Is there an element that does nothing under addition? $-5 + 0 = -5$ and $0 + 3 = 3$. The element 0 is part of the set and under the operation of addition it returns the identical element. 0 is the *additive identity*. If we consider the integers with multiplication, we notice that multiplying by 1 returns the identical value. $-5 \times 1 = -5$ and $1 \times 3 = 3$. 1 is the *multiplicative identity*. It is important to note that if we begin with the set of natural numbers and the operation addition, the identity property does not hold. 0 is usually the additive identity, but 0 is not a natural number. Consequently, there is no additive identity for the natural numbers.

Inverse Property. In order to consider if a set with an operation has the *inverse property*, there must first be an identity element. An inverse element is an element that pairs up to return the identity. Let's use an example: consider the integers under addition. The identity element is 0. Is there a number you can add to 5 to get 0? Yes, $5 + (-5) = 0$. 5 and -5 are *additive inverses*. Every element has an additive inverse, so the inverse property holds for the integers under addition. However, if we consider the integers under multiplication, the inverse property does not hold. 1 is the *multiplicative inverse*. However, is there a number you can multiply to 5 to get 1? You may argue $5 \times \frac{1}{5} = 1$. You are correct. However, $\frac{1}{5}$ is not an integer. If you can think of at least one counter-example, then the inverse property does not hold.

Associative Property. You have often come across the *associative property* in your K-12 curriculum. Symbolically it says that $(a \blacksquare b) \blacksquare c = a \blacksquare (b \blacksquare c)$. In words, the associative property says that we can regroup elements and produce the same result. For example,
$$(2 + 5) + 8 = 7 + 8 = 15 = 2 + 13 = 2 + (5 + 8)$$
However, consider the operation we defined as $a * b = a^b$.
$$(2 * 3) * 4 = 2^3 * 4 = 8 * 4 = 8^4 = 4096$$
But
$$2 * (3 * 4) = 2 * 3^4 = 2 * 81 = 2^{81}$$
$$\approx 2,400,000,000,000,000,000,000,000$$

Abelian Groups. Now let's bring the seven ideas together for a single mathematical construct. An *abelian group* is defined as any *set* with a *closed, binary operation* for which the *identity*, *inverse*, *associative*, and *commutative* properties hold. The set of integers under the operation addition form an infinite abelian group. Abelian groups are named after the 19th century Norwegian mathematician Niels Abel.

Consider the set of least residue system modulo 5 under the operation addition modulo 5. Since the system is finite instead of infinite, we can set up a table to examine all of the results:

+ mod 5	0	1	2	3	4
0	0	1	2	3	4
1	1	2	3	4	0
2	2	3	4	0	1
3	3	4	0	1	2
4	4	0	1	2	3

Do the integers modulo 5 form an abelian group under addition?
- ✓ The set is closed under addition modulo 5. We can see that any of the numbers $\{0, 1, 2, 3, 4\}$ added to any other number result in the numbers $\{0, 1, 2, 3, 4\}$.
- ✓ From the table, we can see that the identity is 0. If we add 0 to any number $\{0, 1, 2, 3, 4\}$, we return the identical number.
- ✓ Does every element have an inverse that returns 0?
$$1 + 4 = 0$$

$$2 + 3 = 0$$
$$3 + 2 = 0$$
$$4 + 1 = 0$$
$$0 + 0 = 0$$

Yes! Every element has an additive inverse, so the inverse property holds.

- ✓ Addition is both commutative and associative.
- ✓ The integers modulo 5 form a finite abelian group under addition.

An abelian group is an example of a strong mathematical structure. Abstract algebra contains many important theorems about abelian groups. In a sense, abelian groups behave nicely. Most of the mathematics you have experienced are based on abelian groups. The difference in this chapter is that you have experienced finite abelian groups: modular systems.

A Non-numerical Example

At this point in the chapter most people wonder what the big deal is. Why the hype over abelian groups? Consider the following example.

I am going to make up a set: $\{\&, ?, @, \%\}$. Now I am going to make up an operation $\$$ that is defined by this table:

$	&	?	@	%
&	%	@	&	?
?	@	%	?	&
@	&	?	@	%
%	?	&	%	@

Does the set $\{\&, ?, @, \%\}$ under the operation $\$$ form an abelian group?

- ✓ The closure property holds as no new elements are introduced in the chart. We begin with $\{\&, ?, @, \%\}$, and the only outcomes are $\{\&, ?, @, \%\}$.
- ✓ The identity property holds. If you look carefully, you can see that $@$ is the identity. Anything that is $\$$ with $@$ returns the identical value.

$$\& \$ @ = \&$$
$$? \$ @ = ?$$
$$@ \$ @ = @$$
$$\% \$ @ = \%$$

- ✓ The inverse property holds. Every element has an inverse element that brings it back to the identity $@$.

$$? \$ \& = @$$
$$\% \$ \% = @$$
$$\& \$? = @$$
$$@ \$ @ = @$$

- ✓ The symmetry down a diagonal line reveals that the operation $\$$ is commutative. Extra checking reveals that $\$$ is associative, too.

So yes, the set $\{\&, ?, @, \%\}$ under the operation $\$$ forms an abelian group.

This non-numerical example demonstrates that mathematical structure can exist independent of numerical values. Many people who first peruse and abstract algebra text ask, "Where are the numbers? I thought this was a math book." As we have seen, mathematical structures do not need to involve infinite sets of numbers. In fact, they do not need to involve numbers at all! This should make us wonder what exactly mathematics *is*.

Something to Think About

The study of modular arithmetic and finite abelian groups encourages us to ask two questions. The first has to do with the nature of mathematics. Do we make up mathematics, or do we discover it? In this chapter, we have seen the integers reorganized into modular systems. We defined a group. We worked through numerical and finally non-numerical examples of abelian groups. Clearly, the use of numbers falls into the world of mathematics. However, abelian groups and the non-numerical examples are also part of the study of mathematics. What is math? Is it something we made up? If so, why did we create it? If not, could we say that God made up abelian groups and we discovered them? On what biblical basis do we answer these questions?

The second consideration does not prompt direct biblical integration; it's actually an analogy. People hold many different beliefs about the nature of time and the universe. Time and reality can be seen as the real number line: eternally continuous with no beginning or end. They could also be understood as a line segment: a clear beginning and a clear end. Another major view is a picture akin to modular arithmetic: reality *is* cycles into which humans are reincarnated to repeat the sequence over and over again. Which mathematical analogy do you think best represents time? Which represents the universe and our reality? Which represents our lives?

Covering the Reading

1-4. Find the least positive residue *modulo 4*.
   1. $3 + 3$
   2. 3×3
   3. 57
   4. 14×53

5. Identify a modular system in modern life not mentioned in this chapter.

6. Do the integers under division have the closure property? Explain.

7. What is the identity element on a clock? Explain.

8. Do the rational numbers under multiplication have the inverse property? Explain.

9. What is the difference between the commutative and associative properties?

Problems

10-12. Which of the following form an abelian group? Provide the necessary rationale. Assume addition is commutative and associative.

10. The rational numbers under addition.

11. The integers under addition modulo 7.

12. The integers under multiplication.

13. Does the set {M, A, T} form a group under the operation H as delineated in the table below? Provide the necessary rationale.

H	M	A	T
M	T	M	A
A	M	A	T
T	A	T	M

14. What is math? Is it something we made up? If so, why did we create it? If not, could we say that God made up abelian groups and we discovered them? On what biblical basis do we answer these questions?

15. Which mathematical analogy would you argue best represents time? Which represent the universe and our reality? Which represents our lives?

Chapter 3.6
Algebraic Reasoning

What *is* algebra? For many people, algebra is the first time that it feels like math ceases to be math. In math class before algebra, we actually do things with numbers. Algebra partially abandons numbers for letters. We begin the quest to find "x" through a variety of methods that we may or may not fully understand. All of these accusations are partially true. However, an closer look at the history of algebra will help us better grasp the beauty and elegance of the subject.

Activity 1
Try to solve each of the following puzzles.
1. Some number's square and 16 is the same as 25. What is the number?
2. When this number's cube is added to this number's square, it is the same as twice the number. What is the number?
3. Given my number, your number is five more than three times my number. What are our numbers?

Historical Context

You may or may not like "word problems," but word problems were the birthplace of algebra. Many modern people cannot imagine a time without the rules of algebra existing in some dusty old parchment somewhere. However, at the time of Diophantus (c. A.D. 250), sometimes called the "Father of

Algebra," mathematicians were just beginning to develop algebra and understand algebraic properties. Diophantus' *Arithmetica* contained many puzzle problems similar to the ones in the warm-up activity.

This practice of solving word problems continued in the history of mathematics, particularly during the so-called Dark Ages of the West, when the intellectual center of mathematics moved east to Baghdad.

In the East, algebra was further developed as a set of tools to uncover or pry open these word puzzles. The ninth-century Arabic mathematician al-Khwarizmi, sometimes called the "Father of Algebra," wrote a treatise summarizing the rules for solving what we would call algebraic problems. He wrote

Hasib al-jabr w'al-muqabala, a book on *al-jabr* (completing or restoring broken parts) and *al-muqabala* (balancing). The word algebra is derived from the word *al-jabr*. The essence of algebra is to restore what is unknown by keeping things balanced.

During the rebirth of mathematics in the West, the sixteenth-century French mathematician François Viète, sometimes called the "Father of Algebra," solidified the use of *variables* to represent values in algebra. He systematized algebra into a set of rules that can be applied to help reveal the unknown. Later mathematicians, none of whom were called the "Father of Algebra," pursued and managed progressively more difficult algebraic relationships. So as you can see, the historical development spans at least 1600 years, growing in time to encompass a progressively rich understanding and systematizing of numerical relationships.

Concept Development 1: Solving Equations

At its heart, elementary algebra is a systematized way to reveal unknown values. Let's also allow ourselves to use the modern notations of algebra such as $2x + 5 = 17$. How would you reveal the unknown value x? Basic problems such as this could be solved by guessing and checking. However, is there a way to systematically find the value for x? Certainly. You remember your rules of algebra. Add -5 to both sides of the equation to reveal that $2x = 12$. Then multiply both sides of the equation by $\frac{1}{2}$ to reveal that $x = 6$. On what basis do these "rules" work? What are the rules? And what is an equation?

http://upload.wikimedia.org/wikipedia/commons/d/d0/
Balance_scale_IMGP9755.jpg

First, an *equation* is simply a balance. The sign "=" means "is the same as." Consider a very simple equation like $2 + 3 = 5$. It states that two added to three is the same as five. Another equation is $8 = 8$. Since an equation is a balance, what can you do to both sides of a balance that does not change its essence? We can add the same thing to both sides of a balance; it is still balanced. We can subtract the same thing from both sides of a balance, and it is still balanced. Can we multiply? Divide? Square? Take a square root? Take an absolute value? Take a logarithm? Take a derivative? What actions maintain a balance?

The process of solving an equation is revealing the unknown through keeping a balance. Algebraic thinking includes this major concept of balance. A common, non-algebraic error in thinking is that we can "move something to the other side" of the equals sign. If we had items on two sides of a balance and they were balanced, then moving something to the other side would leave the situation unbalanced. It would no longer be an equation.

When considering the long history of algebra, people usually ask "If mathematicians knew the general algebra rules in the first few hundred years of the subject's development, what were they trying to do for the next millennia?" Good question. Essentially, they were solving equations *in general terms.*

An example of a specific equation is $2x + (-9) = 0$. We can manipulate the balance through the rules of algebra to reveal the unknown:

Line 1 – given:
$$2x + (-9) = 0$$

Line 2 – add equals to both sides:
$$2x + (-9) + 9 = 0 + 9$$

Line 3 – regroup:
$$2x + (-9 + 9) = 9$$

Line 4 – simplify:
$$2x + 0 = 9$$

Line 5 – simplify:
$$2x = 9$$

Line 6 – divide both sides by equals:
$$\frac{2x}{2} = \frac{9}{2}$$

Line 7 – regroup:
$$\frac{2}{2}x = \frac{9}{2}$$

Line 8 – simplify:

$$1x = \frac{9}{2}$$

Line 9 – simplify:
$$x = \frac{9}{2}$$

However, suppose you were given a more general equation such as $ax + b = 0$ where a and b are integers $\{..., -3, -2, -1, 0, 1, 2, 3, ...\}$. We want to know if there is a solution to this general equation. Using the same manipulations, $= -\frac{b}{a}$. So there does appear to be a solution for this equation, and we can find it by dividing the opposite of the constant by the coefficient of x. The only restrictions are that the answer might be a fraction and $a \neq 0$.

That was not much of a challenge. However, what if the general equation is $ax^2 + b = 0$? Using similar manipulations:
$$ax^2 + b = 0$$
$$ax^2 = -b$$
$$x^2 = -\frac{b}{a}$$
$$x = \pm\sqrt{-\frac{b}{a}}$$

There appear to be two solutions: $x = \sqrt{-\frac{b}{a}}$ and $x = -\sqrt{-\frac{b}{a}}$. However, you should be concerned at this point. What is the square root of $-\frac{b}{a}$? If a and b are both positive or both negative, then the solution is the square root of a negative number, an *imaginary* number. However, if either b or a are negative, then the solutions are real numbers. One small change in our initial equation makes a significant difference in the possibility of a real solution.

As early as Euclid (300 B.C.), mathematicians knew any equation of the form $ax^2 + bx + c = 0$ had a general solution. This equation is the general quadratic (degree 2) equation. You probably memorized and forgot this formula, the quadratic formula: $x = \frac{-b \pm \sqrt{b^2 - 4ac}}{2a}$. You have certainly used it without knowing its significance. This formula finds the solutions, real or imaginary, for *any* quadratic equation! That is pretty important. Any quadratic equation can *easily* be solved by a pre-algebra student. So if these mathematicians in the East and during the Renaissance knew this, what further challenges did they face?

The pursuit of a general solution for a cubic equation, $ax^3 + bx^2 + cx + d = 0$, consumed much of the algebraic efforts of mathematicians in the first four centuries of the last millennium. Eventually, "the cubic" was solved and a formula emerged that shows that given the coefficients, we can solve any equation of degree 3 with real coefficients. Here is the solution to the cubic:

$$x_1 = -\frac{b}{3a}$$
$$- \frac{1}{3a}\sqrt[3]{\frac{1}{2}\left[2b^3 - 9abc + 27a^2d + \sqrt{(2b^3 - 9abc + 27a^2d)^2 - 4(b^2 - 3ac)^3}\right]}$$
$$- \frac{1}{3a}\sqrt[3]{\frac{1}{2}\left[2b^3 - 9abc + 27a^2d - \sqrt{(2b^3 - 9abc + 27a^2d)^2 - 4(b^2 - 3ac)^3}\right]}$$

$$x_2 = -\frac{b}{3a}$$
$$+ \frac{1 + i\sqrt{3}}{6a}\sqrt[3]{\frac{1}{2}\left[2b^3 - 9abc + 27a^2d + \sqrt{(2b^3 - 9abc + 27a^2d)^2 - 4(b^2 - 3ac)^3}\right]}$$
$$+ \frac{1 - i\sqrt{3}}{6a}\sqrt[3]{\frac{1}{2}\left[2b^3 - 9abc + 27a^2d - \sqrt{(2b^3 - 9abc + 27a^2d)^2 - 4(b^2 - 3ac)^3}\right]}$$

$$x_3 = -\frac{b}{3a}$$
$$+ \frac{1 - i\sqrt{3}}{6a}\sqrt[3]{\frac{1}{2}\left[2b^3 - 9abc + 27a^2d + \sqrt{(2b^3 - 9abc + 27a^2d)^2 - 4(b^2 - 3ac)^3}\right]}$$
$$+ \frac{1 + i\sqrt{3}}{6a}\sqrt[3]{\frac{1}{2}\left[2b^3 - 9abc + 27a^2d - \sqrt{(2b^3 - 9abc + 27a^2d)^2 - 4(b^2 - 3ac)^3}\right]}$$

As you are probably asking, why should they stop with degree 3? Mathematicians then pursued degree 4, degree 5, and degree 6 solutions. Mathematicians of the 1500's conquered the degree 4 quartic equation. However, despite the efforts of some of the "greats" of mathematics, no one was able to solve the degree 5 quintic equation or higher. Yes, some could be solved, but remember, our initial question was whether there is a general equation to solve *all* quintic equations. The algebraists of the 19th century, especially the French mathematician Évariste Galois, eventually managed the problem and determined that there can be no general solution to degree 5 or higher polynomials using elementary operations. Galois went on to classify which quintics could and could not be solved before he died at the age of 20 in a dual. So as you can tell, the simple question of solutions to polynomials drove mathematics, specifically algebra, forward for over 1000 years. It turns out that elementary algebra had a lot more to offer then "adding 3 to both sides of the equation."

Concept Development 2- Modeling

Algebraic thinking involves the manipulation of equations. However, algebraic reasoning goes beyond the balancing act. Applied algebra is built on the idea that we can model situations with an equation. Let's take a look at a few models to help us understand what this means.

Activity 2

Write an equation that describes each situation:

1. The cost of a taxi ride is $2.50 plus $1.75 for each mile.
2. Suppose you begin with $3. If I double your amount every day, how much money will you have after d days?
3. The cost of a taxi ride is $2.50 plus $1.75 for each mile or portion thereof.

These three examples are fairly straightforward. Much of the natural world and our human endeavors tend to be far more complicated. As a result, most phenomena we try to model do not come in a straightforward analytical format like the ones you just worked with. In real-world contexts, the application of most algebraic equations begins with a set of data. For example, suppose the numerical data from questions 1 and 3 were presented in a table:

Minutes	0	1	2	3	4	5
Cost	2.50	4.25	6.00	7.75	9.50	11.25

Or suppose the data were plotted graphically on a scatterplot:

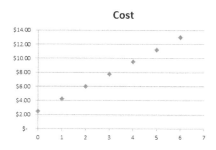

In each case, analytical, numerical, or graphical, the data can be modeled algebraically. Applied algebraic reasoning asks that we attempt to model situations with algebraic equations.

Something to Think About

What is algebra and where does algebra come from? Is it simply a game? It is created by humans? Is it created by God and discovered by humans? Is algebra in some way tied to the natural world? Clearly, it is useful and effective. What makes this the case?

Covering the Reading

1 – 3. Solve the equations in Activity 1 using algebra or other methods.

1. This number's square and 16 is the same as 25. What is the number?

2. When this number's cube is added to this number's square it is the same a twice the number. What is the number?

3. Given my number, your number is five more than three times my number. What are our numbers?

4. Solve $3x^2 + 4x - 10 = 0$ using the quadratic equation.

5. Explain why in algebra we do not "move it to the other side."

6 – 8. Complete Activity 2. Write an equation that describes each situation.

6. The cost of a taxi ride is $2.50 plus $1.75 for each mile.

7. Suppose you begin with $3. If I double your amount every day, how much money will you have after d days?

8. The cost of a taxi ride is $2.50 plus $1.75 for each mile or portion thereof.

Problems

9. Explain what an equation is and what it means to solve an equation.

10. Explain the significance of the quadratic formula. What does it do? Why is it important?

11. What does it mean to "solve the cubic"?

12. Reflect and write on these questions: What is algebra and where does algebra come from? Is it simply a game? It is created by humans? Is it created by God and discovered by humans? Is algebra in some way tied to the natural world? Clearly it is useful and effective. What makes it so?

Chapter 3.7
Combinatorial Reasoning

Although many of you have spent extensive hours in various algebra classes, few can claim to have had a course in combinatorics. Some have had a brief exposure to combinations in the rarely seen chapter 13 of an algebra 2 book, or if the subtleties of a statistics course were grounded in probability distributions. In any case, the American educational system has historically been strong in developing algebraic skills, some Euclidean geometric thinking and proof, and preparing students to move on to briefly grasp the holy calculus grail. However, the system has been deficient in deeply developing the minds of students in other exciting areas of mathematics such as number theory, probability, game theory, fractal geometry, topology, and combinatorics, to name a few. It may be apparent now why this book is so valuable.

Both the familiar and neglected areas of mathematics are accessible to everyone with some basic study, but they take significant effort to master. In this chapter, we will examine the difference between algebraic argument and combinatorial argument (an argument involving combinations). The point is not to downplay the power of algebraic argument but to open your mind to an alternate way of thinking about mathematics. Furthermore, in the following cases, I hope you find the combinatorial arguments in some way more satisfying than the algebraic arguments, maybe motivating you to further study.

Introduction to Combinations

A combination in mathematics is similar to other instances of "combinations" you may be familiar with. Your high school locker may have had a 3-digit combination. It was a collection of 3 numbers, usually from the numbers 1-40, that you used in a particular order. However, the mathematical concept of combination diverges a little from your locker combination. In mathematics, the combination 18-34-12 is the same as the combination 12-18-34. On your lock, the order of the numbers mattered; in mathematical combinations, order does not matter. But mathematics does deal with situations in which order matters; they are called *permutations*. Another important note is that the "combination" 18-34-18 does not qualify as a mathematical combination because a particular number cannot be repeated.

We also need to briefly discuss notation, or how combinations are written down. Given a set of numbers or objects of size n, the total number of different combinations of size r that can be formed from the set of size n is denoted by the function $C(n,r)$. It could also be written $\binom{n}{r}$.

Since combinations and combinatorial argument are probably new to you, you should read this chapter very slowly and carefully.

Warm-up Activity

Suppose you have 6 objects labeled [A,B,C,D,E,F]. How many ways can they be chosen 2 at a time? List the $\binom{6}{2} = 15$ possibilities here: (remember, AB is the same as BA in mathematical combinations, so AB and BA should only be counted as 1).

Now, suppose there 5 people, A, B, C, D, and E, being considered for a 3-person committee. The total number of committees that could be formed (the order of names does not matter) is $\binom{5}{3} = 10$. List the 10 committees here:

Concept Development

To more fully develop our thinking in terms of combinations, we need to take a few moments to investigate the locker type "combinations" in which repeating is allowed—a condition we will eliminate when focusing on mathematical combinations. We must consider two criteria when counting arrangements of items: (1) does order matter? And (2) is there "replacement" - that is, can an element be repeated? These considerations produce a 2x2 chart:

Counting arrangements of n items taken r at a time.	Order matters (AB is different than BA)	Order does NOT matter (AB is the same as BA)
No replacement/ No repeating of elements	1.	2.
Replacement/ Repeating of elements	3.	4.

Mathematical *combinations* meet the requirements for box number 2 and are denoted by nCr, C(n,r), or the notation used in this text: $\binom{n}{r}$. Box number 1 describes what are called *permutations*; these are denoted nPr. Box number 3 is derived directly from the multiplication counting principle below. Now that you have a better handle on the vocabulary, let's examine each box in more depth:

Box 4: Box 4 is a bit more complicated than the focus of this book, so we will no develop it further. For the curious, here is the formula: $\binom{n+r-1}{n-1}$.

Box 3: Box 3 involves choosing r items from a collection of size n where order matters but the n items are always available even after each choice is made. A practical parallel situation is making "words" from our 26 letter alphabet. Suppose we wanted to make 3 letter "words" where a "word" is any 3-letter arrangement. Examples of words include "ABA" and "CFG." Furthermore, "ABA" is a different word from "AAB." The way to find the total number of "words" is to consider the number of options in each position and multiply them: 26 options for the first position x 26 options for the second position x 26 options for the third position. (Multiplication is clearly the operation here: for example, with an "alphabet" of 3 letters (A,B,C) and "words" of length 2, there are 3×3 "words." The letter A is first matched with each of the other three letters. The letter B is then matched with each of the other three letters. Finally, the letter C is matched with each of the other three letters. The total is the result of repeated addition: multiplication. In this case there are 3 groups of 3.) The total number of 3 letter "words" from the 26 letter alphabet is $26 \times 26 \times 26$, or 26^3. Similarly, the number of 5 letter "words" is 26^5. If we used the 24 letter Greek alphabet, then the number of 5 letter "words" is 24^5. We can generalize this result of arranging n items taken r at a time where order matters and replacement is allowed as n^r.

Box 1: Box 1 permutations lead directly to the focus of this chapter: combinations. Consider box 1, which involves the number of ways to arrange r out of n items when there is NO replacement and order DOES matter. The argument is similar to the argument for box 3 except that letters CANNOT be repeated. "ABA" is not a "word" in this case. How many 3 letter "words" are there when letters CANNOT be repeated? Try to answer this question before moving on.

The number of 3 letter "words" if letters CANNOT be repeated can be calculated using the same principle as above. There are 26 options for the first letter. However, once one is used, there are 25 options left for the second letter, then 24 options for the third letter. Therefore, there are 26 x 25 x 24 possible 3 letter "words" under these conditions. The number of 5 letter words would be 26 x 25 x 24 x 23 x 22. The number of 5 letter Greek words would be 24 x 23 x 22 x 21 x 20.

Generalizing this result takes a bit of algebraic manipulation, but it follows directly from the examples above. For sake of simplicity, consider a 10 letter alphabet. How many 3 letter "words" can we generate in the "order matters" but "no repeating" conditions? Obviously, there are 10 x 9 x 8 "words." Generalizing this situation for an n-lettered alphabet and r-letter words is simply n x (n-1) x (n-2) x ... x (n-r+1). However, a little arithmetic produces a simpler formula. Consider that

$$10 \times 9 \times 8 = \frac{10 \times 9 \times 8 \times 7 \times 6 \times 5 \times 4 \times 3 \times 2 \times 1}{7 \times 6 \times 5 \times 4 \times 3 \times 2 \times 1}$$

since the ratio 7x6x5x4x3x2x1 in the numerator and denominator is 1. In mathematics, we use an operation called *factorial* as a shortcut to writing a product like 7x6x5x4x3x2x1. Factorial notation employs an exclamation mark: 7! = 7x6x5x4x3x2x1. So,

$$10 \times 9 \times 8 = \frac{10 \times 9 \times 8 \times 7 \times 6 \times 5 \times 4 \times 3 \times 2 \times 1}{7 \times 6 \times 5 \times 4 \times 3 \times 2 \times 1} = \frac{10!}{7!}$$

This factorial shortcut allows us to generalize the permutations in box "1." To summarize, then, n items taken r at a time when order matters and there is NO replacement is $nPr = \frac{n!}{(n-r)!}$.

Box 2: This chapter focuses on the combinations found in box 2; they can be derived from the permutations formula with one minor modification. As you know, the difference between combinations and permutations is order. In combinations, order does NOT matter. In permutations order does matter. In both cases there is NO replacement. So the question to consider is "how many combinations are there of n items taken r at a time when order does NOT matter and there is NO replacement?"

Suppose the formula for box 2 = nCr = C(n,r) = $\binom{n}{r}$, which we read "n choose r." How would it relate to nPr? Let's consider an example. Suppose we are trying to make 3 letter "words" from a 5 letter alphabet [A,B,C,D,E]. In permutation, the "word" ABC is different from ACB, BAC, BCA, CAB, and CBA. However, in combinations, those 6 "words" are considered the same. Consequently, for *every* combination "word" there are 6 permutation "words." We can also write this as a formula:

$$_5P_3 = \binom{5}{3} \times 6$$

Where did the "6" come from? For nCr we would choose 3 letters, then order them to find the permutations of that choice. Three distinct letters can be organized in 3x2x1 = 3! = 6 different ways. To generalize, if we would choose r letters and order them, the number of orders would be r! for each choice. For the specific example,

$$_5P_3 = \binom{5}{3} \times 3!$$

More generally:

$$nPr = \binom{n}{r} \times r!$$

Which algebraically becomes

$$\binom{n}{r} = \frac{nPr}{r!} = \frac{\frac{n!}{(n-r)!}}{r!} = \frac{n!}{(n-r)!r!}.$$

Now that we have finished our exploration, we can fill in the chart:

Counting arrangements of n items taken r at a time.	Order matters (AB is different than BA)	Order does NOT matter (AB is the same as BA)
No replacement/ No repeating of elements	1. Permutations: $$nPr = \frac{n!}{(n-r)!}$$	2. Combinations: $nCr = C(n,r) =$ $\binom{n}{r} = \frac{n!}{(n-r)!r!}.$
Replacement/ repeating of elements	3. n^r	4. $\binom{n+r-1}{n-1}$

We have developed the basic rules for counting combinations and permutations. This task is analogous to learning how a hammer or saw work. Now we will build some beautiful mathematical thinking.

Combinatorial Argument

What makes one mathematical argument superior to another? Most people involved in mathematics have an intuitive sense of quality proof; there are very few objective criteria. Nonetheless, two criteria are worth considering. One criterion is the simplicity of the argument; does one argument take a more direct route than another? Second, the proof argument is strong when it is clearly or easily meaningfully connected to the problem under consideration.

It is on this second criteria that algebraic argument frequently loses to other arguments. Often, when considering an algebraic argument, the mathematician is frequently able to ignore the meaning of the problem under consideration. Instead, she applies algebraic tools or rules that are powerful and secure while ignoring or veiling meaning.

Algebraic Arguments. Algebraic thinking is different than algebraic argument. Algebraic thinking is usually associated with the ability to recognize and extend or summarize patterns. Algebraic thinking is intentionally imbedded in quality K-12 curricula. For instance, in first grade, my daughter frequently brought home math worksheets that have a specific section indicating "Algebraic Thinking." Obviously, she did not have to sit at home solving equations. Instead, she had to extend patterns and then write a rule for the pattern she has extended. This "algebraic thinking" leads to the familiar linear, quadratic, exponential, and trigonometric equations of high school algebra.

When we use algebra rules associated with the high school courses to make arguments in algebra or other areas of mathematics, we are employing "algebraic argument." While it is often used in combination with other ways of thinking, algebraic argument can also stand on its own. Let's consider two examples to explain how algebraic argument is used in geometry and number theory.

Example 1: The famous Pythagorean Theorem states that in any right triangle, the square of the hypotenuse equals the sum of the squares of the other two sides. Geometrically, what is happening is that literally the area of the square that is made by the length of the hypotenuse is the same as the total area of the two squares formed by the two sides. Mathematicians have written many geometric proofs of the theorem, including Euclid in his *Elements* written c. 300 B.C. However, the Pythagorean theorem can also be proven with the support of algebra. In this case, the goal is to show algebraically that $a^2+b^2=c^2$ (a format which is familiar to most). Here is an example of an algebraic argument.

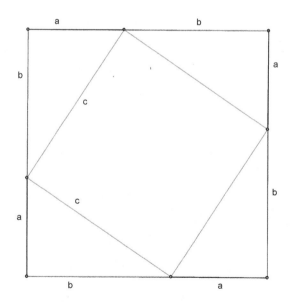

Consider the figure. The total area of the square can be calculated in two ways: (1) the area of the entire square using the outside dimensions, or (2) adding the areas of the five figures inside the square. Taken the first way, the area of the square is (a+b)(a+b). Taken the second way, the area is the sum of four triangles plus the little square in the middle. Since the area of a triangle is ½ base x height, the area of each of the four triangles is ½ a x b. Finally, the area of the little square is c x c = c^2. Our reasoning tells us that these methods are both correct; therefore, both results must be equal.

Algebraically,

$$(1)\ (a+b)(a+b) =$$
$$(½\ ab) + (½\ ab) + (½\ ab) + (½\ ab) + c^2$$

Using algebraic rules of expanding and simplifying:

$$(2)\quad a^2 + ab + ab + b^2 = 4(½ab) + c^2$$
$$(3)\quad a^2 + 2ab + b^2 = 2ab + c^2$$

Subtracting 2ab from each side of the equation:
$$(4)\quad a^2 + b^2 = c^2$$

Notice that once the algebraic equation was set up, the algebraic argument proceeded without needing to understand the geometric meaning of the process involved. For instance, between step 2 and 3, we simplify 4(½ ab) to 2ab. Geometrically, we are maneuvering the 4 triangles into 2 rectangles. That geometric meaning or interpretation is not required for us to proceed through the algebraic argument.

Example 2: A second example may be helpful in understanding what is meant by algebraic argument before considering specific examples involving Combinatoric Argument. This example emerges from basic number theory. First, let's consider a few questions: when we add two even numbers, is the sum even or odd? When we add two odd numbers, is the sum even or odd? When we multiply two even numbers, is the product even or odd? When we multiply two odd numbers, is the product even or odd? We can answer each of these questions with an algebraic argument. Just like the Pythagorean Theorem argument, the algebraic number theory argument begins by explaining the nature of the problem but then moves to a purely algebraic approach before returning to the nature of the problem.

The sum of two evens. We know that every even number can be written as two times some integer (the set of integers: {...-3,-2,-1,0,1,2,3...}) (e.g. $120 = 2 \times 60$ and $38 = 2 \times 19$). This allows us to express two different even numbers as $2n$ and $2m$ where n and m are integers. Algebraically,

(1) $2n + 2m =$
(2) $2(n+m)$

$n + m$ is an integer, so $2(n+m)$ is an even number. The sum of two even numbers is *always* an even number.

The sum of two odds. Every odd number can be written an even number plus one (e.g. $121 = 120 + 1 = 2 \times 60 + 1$ and $39 = 38 + 1 = 2 \times 19 + 1$). We can therefore express two different odd numbers as $2n+1$ and $2m+1$ where n and m are integers. Algebraically,

(1) $(2n+1) + (2m+1) =$
(2) $2n+2m+2 =$
(3) $2(n+m+1)$

$n + m + 1$ is an integer, so $2(n+m+1)$ is an even number. The sum of two odd numbers is *always* an even number.

The product of two evens. Let $2n$ and $2m$ where n and m are integers be two even numbers. Algebraically,

(1) $2n \times 2m =$
(2) $4nm =$
(3) $2(2nm)$

$2nm$ is some integer, so $2(2nm)$ is an even number. The product of two even numbers is *always* an even number.

In these proofs from geometry and number theory, we turned each case into an algebraic description. After converting to algebraic form, we manipulated the equations without regard for the context of the problem. This method is effective and legitimate. However, an argument that clearly maintains the essence and meaning of the problem is more satisfying. We will consider the contrast between algebraically obscured argument and combinatorially meaningful argument next.

Algebraic vs. Combinatorial Argument (Example 1)

Consider the Warm-Up activity in which we listed the number of ways to choose 2 items from a list of 5. There are 10 possibilities. We can apply the combinations formula in this case to find that total: $\binom{n}{r} = \frac{n!}{(n-r)!r!}$ where "!" means factorial. (e.g. $5! = 5 \times 4 \times 3 \times 2 \times 1$).

(1) $\quad C(5,2) = \frac{5!}{(5-2)!2!} =$

(2) $\quad \frac{5!}{(3!)!2!} =$

(3) $\quad \frac{5 \times 4 \times 3 \times 2 \times 1}{(3 \times 2 \times 1)(2 \times 1)} =$

(4) $\quad \frac{5 \times 4}{2 \times 1} =$

(5) $\quad \frac{20}{2} = 10$

Now, suppose we did some investigating and just calculated some combinations. In the process, we find that not only does $\binom{5}{2} = 10$, but also $\binom{5}{3} = 10$. We continue to investigate and find the following equalities:

$$\binom{6}{0}=\binom{6}{6}$$
$$\binom{6}{1}=\binom{6}{5}$$
$$\binom{6}{2}=\binom{6}{4}$$
$$\binom{7}{0}=\binom{7}{7}$$
$$\binom{7}{1}=\binom{7}{6}$$
$$\binom{7}{2}=\binom{7}{5}$$
$$\binom{7}{3}=\binom{7}{4}$$

What should we inductively conjecture from these results?

$$\binom{n}{r}=\binom{n}{n-r}$$

We are doing mathematics: we have found a pattern, we have generalized, and now we need a proof. Let's consider two: an Algebraic proof and a Combinatorial proof.

Algebraic Proof. Based on the algebraic formula for combinations, this proof is fairly straightforward:

$$(1) \quad \binom{n}{r} = \frac{n!}{(n-r)!r!} =$$

A little algebra trick of adding nothing to a term (in this case adding 0 = n-n to r in the denominator) yields:

$$(2) \quad \frac{n!}{(n-r)!(n-n+r)!} =$$

Factoring out -1:

$$(3) \quad \frac{n!}{(n-r)!(n-(n-r))!} =$$

Using the commutative property:

$$(4) \quad \frac{n!}{(n-(n-r))!(n-r)!} = \binom{n}{n-r}.$$

While this proof required very little actual algebra, it is still an algebraic argument for our conjecture. And now that we have verified that it always applies, it is a theorem. The middle of the argument is simply algebraic manipulation; it requires no understanding of combinations. Let's now take a look at the combinatorial argument.

Combinatorial Proof. Because you have likely not thought combinatorially on a regular basis, we are going to begin with a concrete combinatorial example before arguing the general combinatorial case. Consider the following equality, one of the ones we used in the previous argument.

$$\binom{7}{2} = \binom{7}{5}$$

We would like to understand in a non-algebraic way why $\binom{7}{2} = \binom{7}{5}$. In other words, we want to understand why 7 choose 2 is the same as 7 choose 5. Take a moment to try to argue it in your mind. The left side of the equation is looking for the number of ways to make groups of size 2 out of 7 elements. How many are there? How would we find the total? We could simply take the letters A, B, C, D, E, F, and G and exhaustively list the groups of size 2: AB, AC, AD, etc. The right side of the equation is looking for the number of ways to make groups of size 5 out of 7 elements. How many are there? How would we find the total? We could simply take the letters A, B, C, D, E, F, and G and exhaustively list the groups of size 5: ABCDE, ABCDF, ABCDG, etc. We could make these lists, but if we chose this method, we would prove the theorem for only one case. What can we learn here that will help us generalize to all cases?

Consider some of the groups of size 5. When we chose to include ABCDE in a set, which elements are left out? Elements F and G were left out. When we chose ABCDF, the letters E and G are left out. Let's generalize this pattern. Choosing 5 elements from 7 elements to *be in* a group is the same as choosing 2 elements from 7 elements to be *left out* of a group. In other words, $\binom{7}{2}$ is the number of ways of choosing elements to *be in* a group of size 2 as well as the number of elements to be *left out* of a group when choosing 5 elements. Choosing 5 out of 7 to be in a group is *by default* choosing 2 out of 7 to be left out of the group. (This situation kind of reminds me of when we picked teams on the playground: Captain Brian *not* choosing me to be on his team is ultimately understood as Brian choosing to place me on Captain Wendell's team.) To state it simply, choosing 5 people to be in is choosing 2 people to be out. Now that we have thought through this equation, it should be no surprise that $\binom{7}{2}=\binom{7}{5}$. You just used combinatorial reasoning—granted, on a small scale, to solve this problem. Well done!

We are now prepared to generate a combinatorial argument for our conjecture. To state the problem plainly again, we are seeking to show that $\binom{n}{r}=\binom{n}{n-r}$.

The Proof: Given n items, choosing r items to be in a set by default leaves n-r items out. Therefore, every time a set of size r is made, a set of size n-r is also made. Consequently, the number of ways of choosing r items for a set is the same as the number of ways of choosing n-r for a set. Therefore, $\binom{n}{r}=\binom{n}{n-r}$.

This argument uses no algebra. Instead, it engages the mind to think about what the equality $\binom{n}{r}=\binom{n}{n-r}$ actually *means*. Let's now consider a second but more complex comparison.

Algebraic vs. Combinatorial Argument (Example 2)
As we continue to study combinations, we find another pattern:

$$\binom{6}{4} = \binom{5}{4}+\binom{5}{3}$$

$$\binom{7}{3} = \binom{6}{3}+\binom{6}{2}$$

$$\binom{3}{1} = \binom{2}{1}+\binom{2}{0}$$

$$\binom{5}{4} = \binom{4}{4}+\binom{4}{3}$$

We then conjecture and need to prove that

$$\binom{n}{r} = \binom{n-1}{r}+\binom{n-1}{r-1}.$$

Algebraically, we can prove this by using the definition of combination. However, doing so is not as intuitively satisfying using the combinatorial argument. the combinatorial argument will also allow us to develop a new way of thinking. The most important piece to remember in the algebraic argument is the behavior of factorial (!). Recall that 5! = 5x4x3x2x1, so 5! could be rewritten as 5 x 4! = 5 x 4x3x2x1. In general terms, n! = n (n-1)!.

Algebraic Argument that $\binom{n}{r} = \binom{n-1}{r} + \binom{n-1}{r-1}$.

Start with the right side of the identity:

(1) $\binom{n-1}{r} + \binom{n-1}{r-1} = \dfrac{(n-1)!}{(n-1-r)!(r)!} + \dfrac{(n-1)!}{((n-1)-(r-1))!(r-1)!} =$

(2) $\dfrac{(n-1)!}{(n-r-1)!(r)!} + \dfrac{(n-1)!}{(n-r)!(r-1)!} =$

We know that to combine terms, we must find a common denominator. Expanding a little reveals the fraction that each term needs to be multiplied by to get the common denominator:

(3) $\dfrac{(n-1)!}{(n-r-1)!r(r-1)!} + \dfrac{(n-1)!}{(n-r)(n-r-1)!(r-1)!} =$

Looking at equation (3), it appears that the left term needs to be multiplied by $\dfrac{n-r}{n-r}$ and the right term by $\dfrac{r}{r}$ in order for the fractions to have a common denominator of (n-r)(n-r-1)!r(r-1)! = (n-r)!r!.

(4) $\dfrac{n-r}{n-r} \times \dfrac{(n-1)!}{(n-r-1)!r(r-1)!} + \dfrac{r}{r} \times \dfrac{(n-1)!}{(n-r)(n-r-1)!(r-1)!} =$

(5) $\dfrac{(n-r)(n-1)!}{(n-r)!r!} + \dfrac{r(n-1)!}{(n-r)!r!} =$

(6) $\dfrac{(n-r)(n-1)!+r(n-1)!}{(n-r)!r!} =$

Factoring an (n-1)! from each term in the numerator yields

(7) $\dfrac{(n-1)!((n-r)+r)!}{(n-r)!r!} =$

which simplifies to

(8) $\frac{n!}{(n-r)!r!} = \binom{n}{r}$.

And there we are! We used algebraic argument to conclude that
$$\binom{n}{r} = \binom{n-1}{r} + \binom{n-1}{r-1}.$$

Combinatorial Argument. Let's now argue combinatorially that $\binom{n}{r} = \binom{n-1}{r} + \binom{n-1}{r-1}$. This requires a little bit of thinking about what each of the three terms *means* and how we can use what they mean to build a convincing argument that the identity is always true. You may want to read this slowly and carefully if you are not accustomed to this way of thinking.

First, consider the left side of the identity. $\binom{n}{r}$ counts the number of ways we can make groups of size r out of n people. The right side of the identity must simply be an alternative way of accomplishing the same task: counting the number of ways we can make groups of size r out of n people.

Here is the alternate way to count the groups. Suppose we have n people and want to make groups of size r. Choose an individual from among all the people. For simplicity's sake, let's name him Gauss. We are going to now consider 2 cases: (1) the number of groups of size r that Gauss is in and (2) the number of groups of size r that Gauss is *not* in. By adding these two cases together, we will have counted all of the groups of size r out of n people: those with Gauss in and those with Gauss out.

Counting the two cases reveals the following. First, when counting the number of groups of which Gauss is a part, we have to realize that if we put him into the group, then we need r-1 more people. With Gauss already in a group, there are n-1 people left to choose from. Therefore, with Gauss included in a group of size r, there are $\binom{n-1}{r-1}$ ways to fill up the group of size r.

Now let's consider the case of Gauss being excluded from the group of size r from the start. In this case, we still need r people. However, there are only n-1 left to choose from since Gauss is excluded. Consequently, with Gauss excluded from the group of size r, there are $\binom{n-1}{r}$ ways to make groups of size r.

All we need to do now is add the number of groups that exclude Gauss (all $\binom{n-1}{r}$ of them) to the number of groups that include Gauss (the $\binom{n-1}{r-1}$ of them). The sum = $\binom{n-1}{r} + \binom{n-1}{r-1}$ is the total number of ways to construct groups of size r out of n people = $\binom{n}{r}$.

And there you have it. We just used a combinatorial argument, an argument that relates to the meaning of the combinations involved, to prove that $\binom{n}{r} = \binom{n-1}{r} + \binom{n-1}{r-1}$.

Something to Think About

This study in combinatorics and reasoning prompts us to reflect on two ideas. First, mathematics is quite a diverse subject containing many topics and subtopics. Combinatorics is just one part of mathematics that we barely consider in the K-12 curriculum. Many other really interesting topics like game theory, knot theory, and group theory are also bypassed in regular math classes. Take a moment to google "mathematics topics" and consider the list. Most likely, you have heard of very few of them. One reason this is true is that some topics require the current K-12 mathematics (algebra-geometry track) as a foundation for further exploration. But sadly, many others topics are accessible to K-12 student but simply not considered. Why do we have such a narrow focus in the K-12 curriculum?

A second issue related to combinations is also worth considering. In 1653, Blaise Pascal, a budding mathematician, wrote a treatise on what was later called *Pascal's Triangle*. This book was primarily about combinations. One year later, after a horse accident, Pascal converted to Christianity and left the world of professional mathematics. Basically, his devotion to God motivated him to stop doing mathematics. Is it possible to be a faithful follower of Christ and be a mathematician? Should a Christian be a mathematician or is it a waste of his or her life?

Covering the Reading

1. Complete the Warm-Up activities.

2. Explain the difference between *permutations* and *combinations*.

3. Algebraically, show that the product of two odd numbers is an odd number.

4. What are the strengths and weaknesses of algebraic proof?

5. Summarize the combinatorial proof that
$$\binom{n}{r} = \binom{n}{n-r}.$$

6. Summarize the combinatorial proof that
$$\binom{n}{r} = \binom{n-1}{r} + \binom{n-1}{r-1}.$$

Problems

7. Suppose there are five people and five chairs. In how many different ways can they be seated in the five chairs?

8. Suppose there are twenty players on a baseball team. In how many different ways can a starting lineup (9 players) be constructed if batting order is considered?

9. Suppose there are twenty players on a baseball team. In how many different ways can a starting lineup (9 players) be constructed if batting order is not considered?

10. Notice that $\binom{3}{3} + \binom{4}{3} + \binom{5}{3} = \binom{6}{4}$. Show that this relationship is true algebraically. Can you show the same thing combinatorially? Use combinations and "choosing" sets of people from groups of people to argue that the following is true: The number of ways to choose 3 people from a group of 3 **plus** the number of ways to choose 3 people from a group of 4 **plus** the number of ways to choose 3 people from a group of 5 **is the same as** the number of ways to choose 4 people from a group of 6.

11. Combinatorics is only one of many other topics and subtopics in mathematics that are barely considered in the K-12 curriculum. Why do we have such a narrow (arithmetic-algebra-Euclidean geometry) focus in the K-12 curriculum? Is this approach biblical?

12. In 1653 Blaise Pascal wrote a book about combinations. One year later Pascal converted to Christianity and left the world of mathematics. Is it possible to be a faithful follower of Christ and be a mathematician? Should a Christian be a mathematician or is it a waste of his or her life? Explain and defend your position.

Chapter ∞
The Beginning

I trust that you know by now why I have titled this last chapter "The Beginning." We have explored a wide variety of mathematical topics but have only begun to experience the whole discipline. We have traveled through the mathematical landscape but have stopped at very few of the highlights.

While exploring, we took time to pause in places that prompted answers to the Enduring Questions. A deeper understanding of mathematics demands that we wrestle with philosophical questions regarding the nature of reality, the discernment of truth, and the determination of beauty and goodness. I trust that you will now be able to better answer the questions about what God and your faith have to do with mathematics.

I hope you were not looking for "answers in the back of the book" for these Enduring Questions. They have endured through generations because they are complex. This book has prompted you to begin the process of thinking about them.

In the preface I expressed two goals for the reader: to better understand the breadth of the mathematical landscape and to wrestle with the depth of the interaction between mathematics and the Enduring Questions. These two goals led us to a deeper understanding of our Creator and our place in the universe. However, there is a third goal that I did not reveal at the start of the book.

It is my great desire that you *like* mathematics more after reading this book than you did at the start. I do not know where you began the journey, but I hope you moved toward a greater appreciation for the subject. I do not want to be unreasonable and think that everyone who has read this book now adores mathematics. However, if you hated it, I hope you now tolerate it. If you tolerated it, I hope you now appreciate it. If you appreciated it, I hope you are closer to loving it. Whatever your current status, I hope you want to study more mathematics.

Why study math? Jesus Christ compels us to love God and love our neighbor. A meaningful study of pure and applied mathematics allows us to do both more fully. I trust that this text has developed your appreciation for the vastness of God's work in nature and through humanity. I also hope that this text has moved you to study more mathematics and that, through that study, you will become more effective in serving others.

Enjoy the rest of the journey.

About the Author

Jason VanBilliard has been professionally involved in the field of mathematics education since 1997. His experiences include teaching middle school and high school mathematics, teaching mathematics and pedagogy courses at the undergraduate and graduate levels, serving as a private consultant and a consultant for College Board, working for Educational Testing Service in a variety of contexts, and presenting at local and regional mathematics education conferences. Jason holds a B.S. in 7-12 Mathematics Education and a B.S. in Bible from Cairn University in Langhorne, PA, a M.A. in Mathematics from West Chester University in West Chester, PA, and an Ed.D. in Curriculum, Instruction, and Technology in Mathematics Education from Temple University in Philadelphia, PA.

Made in United States
North Haven, CT
19 November 2022